WHEN GOOD SCIENCE WON

(but it wasn't easy):

CALIFORNIA'S RISE TO EARTHQUAKE SAFETY LEADERSHIP

ROBERT A. OLSON

Whispering Oak Publishing
Orangevale, CA
roal@comcast.net

ISBN: 978-1-7371663-0-6

Printed in the United States

DISCLAIMER: While I benefitted from close relationships with Senator Alfred Alquist, Steve Larson, Karl Steinbrugge, and many others, this is my work. I accept full responsibility for it and hope that people who are committed to saving lives and protecting property benefit from my experiences and observations to further the cause of earthquake safety.

Dedication

I dedicate this book to William A. (Bill) Anderson, PhD, a colleague and friend, who left us prematurely but whose commitment to social science research truly endures as our citizens and communities deal with the threats posed by earthquakes and other hazards. With his encouragement and support, the research was supported by National Science Foundation Grant CMS 9814239.

CONTENTS

Acknowledgments

I learned from and depended on many people to complete this work. Chief among them was my political scientist brother, Richard, Director of the International Hurricane Center and Director of the Extreme Events Institute at Florida International University. I also acknowledge Roger Peterson, who joined with me in the latter stages as a development editor and writing coach. Credit also goes to early advisors that helped conceive and launch this project: William Holmes, structural engineer, and Elliott Mittler, policy researcher.

Along the way I benefitted greatly from the advice and assistance of Fred Turner, structural engineer for the California Seismic Safety Commission; Charles James, former Chief Librarian for UC Berkeley's Earthquake Engineering Center Library; Steve Larson, formerly of California State Senator Alfred Alquist's office; and the staffs of the State Archives, Law Library, and California History Collection. Special credit also goes to Ali Johnston, friend and longtime word processing helper who has seen me through many writings. And always there with her ideas, advice, support, and love was my wife (and former office manager), Natalie.

A more detailed personal description of my key relationships with earthquake engineers, earth scientists, and policy scientists—many of whom became mentors and friends— will be contained in my oral history, which, at the time of this writing, will be published by the Earthquake Engineering Research Institute (EERI).

Foreword

Even when trying to do "the right thing"—protecting Californians from deaths, injuries, and economic losses caused by earthquakes—seismic safety policy advocates have labored in a complex and frequently changing political and governmental context that continues to influence what has been and can be accomplished. Although the stage was set before the March 10, 1933, Long Beach earthquake, state policy history begins a few days after it, including the landmark work of the legislature's Joint Committee on Seismic Safety (JCSS) between 1970 and 1974, and includes the politics of creating California's Seismic Safety Commission, which put earthquake safety on the state policy agenda.

This book tells the stories of who from the JCSS was involved and how several important laws were enacted to help achieve greater safety. Perhaps the book helps dispel a widely held myth, especially outside of California, that its quest for improved seismic safety was "easy." I hope it further empowers Californians and others in earthquake safety (and other in hazardous areas) to learn from our experiences so they can accelerate their efforts to further protect themselves. I also hope this book adds another dimension to public policy studies by documenting the political histories of legislation in a way never before attempted for this specialized field.

I adopted the idea of "policy waves" from a Syracuse University political scientist, W. Henry Lambright, who found the concept useful when studying the long-term evolution of public policy in general. Drawing on historical materials, Wave One sets the stage for state policy actions and examines how two laws following the March 10, 1933, Long Beach earthquake were enacted during the "window of opportunity" that opened immediately afterward. These landmark laws were the Safety of Design and Construction of Public Schools Buildings Act of 1933 (a.k.a. the "Field Act") and the California Earthquake Protection Law of 1934 (a.k.a. the "Riley Act").

Representative government—and the desire to be re-elected—usually means that elected officials, especially legislators, become sponsors of and spokespersons for and against various public policy issues. The content of a legislative proposal depends greatly on who is sponsoring it ("setting the agenda"), and while a legislator may have some particular interests of his or her own, he or she often depends on organized interests to assist with the details and to provide active support or opposition to the subject being considered.

Wave Two discusses the politics of enacting three more laws following the February 9, 1971, San Fernando earthquake. They include laws requiring that local land-use general plans to include a "Seismic Safety Element," that a Seismic Safety Commission be established to carry on the Joint Committee's work after 1974, and that new hospitals must conform to special design and construction standards. Other post-San Fernando Wave Two laws are summarized including ensuring construction in active fault zones to be closely governed and that a state "strong motion instrumentation" program be established to record individual buildings' structural and free field sites' responses to ground motions. I also include a brief chapter about the Joint Committee's role in helping to initiate the political process of restoring California's State Capitol building to resist earthquakes.

Waves Three and Four were the 1989 Loma Prieta and 1994 Northridge earthquakes. I was not involved directly in the legislative responses to those events. Former Seismic Safety Commissioner Lloyd S. Cluff noted that 443 legislative bills addressing seismic safety were introduced following the 1989

Loma Prieta earthquake. The 1994 Northridge earthquake assuredly saw a similar legislative wave. A Senate Office of Research report listed 31 bills that were enacted into law during the 1993–1994, 1995–1996, 1997–1998, and 1999–2000 sessions.

Researchers and practitioners know how important "trigger events" are in creating, expanding, or modifying laws, regulations, and programs by creating "windows of opportunity" for change. The previously cited earthquakes are typical examples. The roles of "issue champions," such as Senator Alfred Alquist, are essential to policy development and implementation. Such leaders become subject matter experts, and other legislators often (but not always) defer to such leaders. Lastly, trigger events can change political agendas by introducing subjects for consideration that are new and demand attention. For example, the original Joint Committee was given only about $10,000 of the legislature's funds to do its work, but immediately after the San Fernando earthquake the president of the Senate offered Senator Alquist whatever funds he needed as responding to the event raced to the top of the legislature's agenda.

It also is valuable to think of the Joint Committee's some 70 volunteer advisors as an "advocacy coalition." Discussed more fully in *Some Buildings Just Can't Dance,* such a coalition is a group of people that comes together to affect basic policy, programs, and governing institutions to achieve the coalition's goals over time. The San Fernando earthquake converted what was to be a study group to an advocacy coalition.

While the use of "waves" illustrates the connection between events and actions, the trigger events themselves have to be politically salient. They must be destructive enough to earn a place on policymaking agendas. For example, the 1952 Arvin-Tehachapi (Kern County) earthquakes, while of sufficient magnitude to cause extensive local damage (especially to unreinforced masonry brick buildings), did not cause any state legislative action. Examples also include the 1951 Eureka, 1957 Daly City, 1969 Santa Rosa, 1973 Point Mugu, and 1980 Livermore quakes, and some others. However, some legislative actions were taken following other relatively minor "wavelets," such as the 1978 Santa Barbara, 1979 Gilroy-Hollister, 1980 Eureka, 1983 Coalinga, 1984 Morgan Hill, and 1987 Whittier Narrows earthquakes.

Although no state legislative actions were taken following the 1975 Oroville event, it did trigger serious local debates about what to do regarding downtown Oroville's damaged brick buildings. Even though a former mayor labeled the battered buildings as "standing rubble," no state seismic policy actions were taken to address the problem posed by thousands of similar unreinforced masonry buildings in the state.

Wave One talks about two stage-setting earthquakes (1925, 1933) and the emergence of advocates who launched the movement for state seismic safety policy.

Part I. Wave One

WAVE ONE, CHAPTER 1:
Two Disasters and New Advocates
Set the Stage

The year 1925, in my opinion, marks the real beginning of earthquake engineering studies and research in the United States. From that year on, the interest and accomplishments in connection with earthquake engineering studies advanced at an accelerated pace.[1]

What made state seismic safety policy action possible immediately following the March 10, 1933, Long Beach earthquake? Partially, it was the legacies of the 1925 Santa Barbara earthquake, but also influential were the non-earthquake caused failure of the newly constructed St. Francis Dam near present-day Santa Clarita in Los Angeles County on March 28, 1928, and the growing professional consensus and widespread public acceptance that California was at risk from future earthquakes.

One researcher observed that:

> in the aftermath of the Long Beach earthquake, seismologists, engineers, and architects promptly and energetically asserted the need for greater seismic safety as the earthquake's main lesson. Repeated again and again in a variety of settings, this message reached a wide audience, and it drew a number of influential endorsements from newspapers and public officials. In addition to shaping public opinion, the campaigners for

3

greater seismic safety also succeeded in prodding the public into action. Under the guidance of seismologists, engineers, and architects, a number of local jurisdictions as well as the state government began to incorporate provisions for earthquake-resistant construction into their building laws.[2]

Santa Barbara Earthquake Provides the Foundation

On June 29, 1925, at about 6:45 a.m. (fortunately before most commercial buildings were occupied), the coastal city of Santa Barbara experienced a damaging earthquake. Excluding damages to dwellings, it was estimated that the event resulted in about $6,230,000 in losses. The earthquake was thoroughly investigated by earth scientists, engineers, and others and reported on in an early (December 1925) issue of the *Bulletin of the Seismological Society of America*. In their paper titled "Earthquake Damage to Buildings," Henry D. Dewell and Bailey Willis, Northern California engineers, noted that:

> In consequence of the clear understanding with which the city manager and his aides grasped the situation, immediately after the earthquake a board of engineers, comprising the most experienced of the profession, was organized to examine the actual conditions and report the facts.
>
> Building operations had apparently been but little controlled, previous to the earthquake, by municipal ordinance or municipal supervision of design and construction.
>
> Public buildings, schools, churches, courthouse, jail, and public library were seriously damaged. Not only schools but hospitals also showed that conscientious work or consideration for the lives of those who might be exposed to danger in the buildings had played but little part in their erection.[3]

The Report of the Engineering Committee on the Santa Barbara Earthquake, dated July 3, 1925, noted:

4

Our inspection of damaged buildings leads us to the conclusion that much of the damage might have been avoided if better material and workmanship had entered into the construction of the buildings affected by the earthquake. Wherever good materials have been properly used and the design has been an intelligent one, slight damage or none has resulted.

We can therefore recommend only that the greatest care be taken as your work of reconstruction is begun [and] that materials and workmanship, coupled with first-class design, be of the best.

In addition, you should add to your building department competent engineers for checking plans and thoroughly reliable inspectors to inspect all work in the course of construction. Whenever the services of an architect or engineer are retained for the getting out of plans, he should also be retained to oversee the carrying-out of these plans in accordance with his specifications in order that in any case of failure of a structure, the blame may be placed where it belongs.[4]

Several policy-relevant activities were initiated after this earthquake. As noted, the City of Santa Barbara called upon the engineering community to analyze and report on the losses. This report quoted above apparently extended to helping write earthquake design provisions for the city's building code. The City of Palo Alto, in Northern California, soon afterward adopted the same provisions. These were among California's first localities to formally require earthquake hazards to be recognized in the design and construction of new buildings.

In addition, the Santa Barbara earthquake led to recommended earthquake design requirements for inclusion in local building codes. These provisions were contained in an appended chapter of the model Uniform Building Code published by the Pacific Coast Building Officials Conference (which became the International Conference of Building Officials, "ICBO"). Although the earthquake provisions were

5

for optional use, according to F. M. Andrus, then an engineer with the County of Los Angeles, "the requirements in the 1927 Edition of this Code are among the first earthquake provisions to be written into any widely used building code in this country."[5] The process accelerated after the 1933 Long Beach earthquake.

The Santa Barbara earthquake also generated an "insurance crisis," which led the State Chamber of Commerce to form a "Committee of 100." It was charged with preparing recommended earthquake provisions for use in building codes. It began its work in 1928, and the committee was still working when the Long Beach earthquake struck on March 10, 1933. The resulting document, *Building Code for California*, not published until 1939, noted:

> The earthquake of 1925 which centered at Santa Barbara, and caused heavy building damage there, resulted in a sudden and wide-spread demand for earthquake insurance. This in turn was followed immediately by an increase in earthquake insurance rates and in the amount of such insurance required by the State Corporation Commissioner before he would approve bond issues on certain types of buildings.
>
> The result of this twofold handicap was ... a sharp recession of building in the state, and in the emergency the aid of the State Chamber was urgently sought by business interests generally. The Chamber undertook an investigation which at once resulted in:
>
> 1. Reduction of earthquake coverage requirements by the State Corporation Commission.
> 2. Reclassification of buildings by the Board of Fire Underwriters and a readjustment of earthquake insurance rates.

But it soon became evident that the fundamental need was for a statewide building code which would require adequate standards of building construction and a reasonable resistance of such construction against earth movements, and to this task the State Chamber directed its attention. This was the genesis of the chamber's sponsorship of a building code.[6]

Next, and thanks to successful lobbying by influential Californians, likely in response to the great 1923 Kanto, Japan, earthquake and fires, Congress in January 1925 authorized the U.S. Coast and Geodetic Survey's (C&GS) Seismological Field Survey to investigate and report on earthquakes. In the summer of 1932, the C&GS had installed three of the nation's first strong motion instruments in the Los Angeles area. One was in Long Beach "almost on top of the earthquake," and "they produced very satisfactory graphs."[7] In fact, this record was the first of its kind ever obtained [in the U.S.] so near the epicenter of a destructive earthquake."[8] The May 18, 1940, El Centro earthquake provided "the best" strong motion record for the time, which earthquake engineers understandably referred to for decades. Far more instruments were needed, and it took the February 9, 1971, San Fernando earthquake to legislatively create California's Strong Motion Instrumentation Program (SMIP).

We can only speculate about how state policy would or would not have emerged had the March 10, 1933, Long Beach not occurred, but it did, and:

> The response to the Long Beach earthquake marked a significant change in Californians' public attitude toward earthquakes. For the first time, a large number of influential newspapers and public officials accepted the contention of seismologists and engineers that Californians needed to guard themselves against seismic hazards. The state government as well as a number of municipalities also for the first time required earthquake-resistant construction; moreover, a group of engineers and bureaucrats with a vested interest in ensuring that buildings were actually earthquake resistant became entrenched within the State Division of Architecture. These developments established a pattern for Californians' subsequent approach to earthquake hazards. The Field and Riley Acts, in much amended form, are still law in California;

7

they have been joined by a number of other acts requiring attention to seismic hazards in the siting and construction of buildings and other structures.[9]

Moreover, the event:

> ... will be less remembered by reasons of its contributions to seismology—for as a crustal tremor there was nothing special about it—than it will be for having broken down the "hush-hush" policy that has hitherto been followed by the commercial organizations of the cities of Southern California.[10]

Dam Failure and the Policy Legacies of California's Dam Act of 1929

People were told that the newly constructed St. Francis Dam, 42 miles north of Los Angeles, was "the safest dam ever built" and "it was the last word in engineering." But "no one told those sleeping in the fertile valley below the dam [at 11:50 p.m. on March 12, 1928]. There wasn't enough time" as telegraph machines clicked "The dam has broken! The St. Francis is out!"[11] The principal cause of the failure was a massive landslide into the reservoir, which was being filled for the first time.

At least 500 people were killed by 36,000 acre feet (about 12 billion gallons) of onrushing water and debris, including timber, bridges, roads, automobiles on the roads, railroad cars and tracks, boulders, animals, barns, homes, and people. The estimated maximum height of the water was 125 feet, and it came down a narrow canyon ultimately covering 65 miles from the dam's site to the coast near Ventura. Devastated communities included Piru, Fillmore, Bardsdale, Santa Paula, Saticoy, and Montalvo. Television's History Channel called the sudden loss of the dam "America's worst civil engineering failure of the 20th century."

California's Governor C. C. Young arrived shortly after dawn on March 12, and President Coolidge "wired that the

8

nation was stunned by the disaster and added that every effort would be made by the federal government to aid state and local agencies in restoration projects."[12] St. Francis Dam's catastrophic failure was a classic "triggering event" that opened the also classic "window of opportunity" for public policy change.

Political Mobilization

The disaster set city, county, state of California, and federal government investigations in motion. A coroner's jury "returned the verdict of 'no evidence of criminal act or intent' on the part of the Board of Water Works and Supply of the City of Los Angeles, or any engineers or employees in the construction or operation of the St. Francis Dam.'"[13]

Within a few months California's legislature dealt with four draft legislative proposals to "prevent repetition of the St. Francis dam disaster" by placing about 600 non-federal dams in California under control of the state government. Interestingly, all four proposals were patterned after an existing state of Pennsylvania statute. One was drafted by State Engineer Edward Hyatt Jr. on behalf of Governor Young's administration, a second by Senator Walter Duvall of Santa Paula, the third by Assemblyman Dan E. Williams of Jacksonville, and the fourth was "arranged by the Farm Bureau Federation."[14]

Following an evening meeting on January 14, 1929, involving Hyatt, Duval, and Williams, it was announced to the press that the four proposals would be consolidated into one, and that an identical version of each (i.e., "companion" bills) would be introduced into the Senate and the Assembly so both houses could consider the same legislation simultaneously. A *Sacramento Bee* newspaper article noted that:

> Jurisdiction over the dams is vested in the state engineering department, which is authorized to make an inventory of

all non-federal dams and reservoirs. The engineer also is directed to supervise the operation and maintenance of the structures, and also the construction of new projects.

Wide police power is conferred upon the engineer in the enforcement of the act, and obstruction of him or any of his agents is made a felony.

For purposes of administering the act, a revolving fund of $250,000 is created, but fees will be charged for supervising the dams.[15]

Processing the Legislation

The formal introduction of Senate Bill 723 by Senators Duval and Mueller (who joined Duval as a coauthor) occurred on January 18, 1929. There is no further mention of a separate Assembly Bill, but it may have been unnecessary given the speed at which SB 723 moved through the political process. The fact that several bills had been combined into one meant that strong consensus existed between Governor Young's administration, the legislative sponsors, and at least the farm bureau lobby.

The Senate's Committee on Governmental Efficiency clearly was where SB 723's substantive issues were managed. Between February 20 and March 28, 1929, the legislation was heard and amended three times. Finally, on March 28 it received a "do pass as amended" recommendation, whereupon SB 723 made its second committee stop: the Senate Committee on Finance.

Finance passed SB 723 out of committee "without recommendation" on April 2. It moved to the Senate floor, where SB 723 was amended three more times while it was on the floor agenda. On May 1, 1929, the Senate approved the bill and sent it to the Assembly for consideration.

SB 723 moved very quickly through the Assembly without any apparent amendments. On May 1 the legislation was referred to the Assembly's Committee on Governmental Efficiency and Economy, and on May 6 it received a "do pass" committee recommendation. Its second stop was the Assembly's Ways and Means Committee, which gave it a "do pass" recommendation on May 10.

SB 723 passed the Assembly on May 14, and it was returned the same day to the Senate for "concurrence." Concurrence was achieved on the same day, and the legislature sent this landmark legislation to Governor Young on May 15, 1929. He signed the new Dam Act on June 10, 1929, stating that:

> This bill provides for a centralized state supervision over dams in California, as to their design, construction and maintenance. In the past an unsatisfactory condition has existed due to the fact that supervision was not thorough or complete [by the state railroad commission].
>
> The present bill concentrates authority in one office, that of the state engineer, who is given authority to do all things necessary to see that dams in California are safe. The bill makes possible the storage of water for irrigation and assures the public that this water is being stored behind safe structures.[16]

The original Dam Act provided $200,000 so a survey could be made of all existing dams in California. A *Sacramento Bee* article noted that "if any of these structures are found to be faulty, the state engineer is empowered to compel the owners to make such changes and reinforcements as to assure their safety."[17]

The Dam Act's Enduring Legacies

The 1929 act institutionalized a strong regulatory role for state government over non-federal dams in California. It not only addressed new dams, but it also called for the evaluation and upgrading or replacement of existing ones—always a difficult political challenge because of American society's general reluctance to "change the rules after the game has been played." In addition, the Dam Act required the state to greatly increase its technical competence so it would have the independent capabilities to review and approve design plans, actual construction, and the long-term maintenance of dams.

The Dam Act of 1929 became the "spiritual centerpiece" of many of California's later state seismic safety laws: the Safety of Design and Construction of Public School Buildings Act of 1933 (the "Field Act"), the dam failure inundation mapping and evacuation planning law of 1972, the Hospital Seismic Safety Act of 1973, and the Essential Services Buildings Seismic Safety Act of 1986. The Dam Act's landmark principles that were brought forward into these later laws include state preemption (or nearly so) of the policy issue in question; development of more stringent building standards and regulations than those administered generally by local governments; and administrative processes to assure that qualified people independently review design plans, specifications, construction drawings, and actual construction. In 1928 no one could have foretold the policy legacies of the St. Francis Dam's catastrophic failure. It also is clear from a later perspective that the 1933 Long Beach, 1971 San Fernando, and 1985 Mexico City earthquakes contributed to the passage of each of the laws discussed above by also opening windows of opportunity.

Emerging Professional Consensus and Advocacy

We must not underestimate the importance of the emergence of a professional consensus about California's risk and the early leadership that addressed it. Others (Geschwind; Meltsner; Olson, Olson, and Gawronski) have written about the earlier lack of scientific agreement about the state's risk and the repression of information and the suppression of activities associated with California's earthquake hazards.

Perhaps one of the best characterizations of this attitude was captured by political scientist Arnold Meltsner when he quoted a 1927 letter to Caltech President Robert A. Millikan (who was to emerge just six years later [1933] as a major Southern California seismic safety leader). The letter from Henry M. Robinson, a bank president and Caltech trustee, urged Millikan "to stop the talk about the immediate approach of an earthquake" by geology Professor John P. Buwalda of Caltech and Harry O. Wood, a research associate of the Carnegie Institute of Technology in Washington, D.C., and a supervisor of what later became Caltech's Seismological Laboratory. Robinson's letter said:

> I wonder if you have any idea how much damage this loose talk of these two men is doing to the values in Southern California. I wonder if you appreciate that one of the effects of the operation of these two men in their wisdom will be to turn the hands of all businessmen against the Institute for bringing into the community men who can talk so glibly about things which they cannot know and which will destroy values unwarrantedly in this whole area. You can hardly appreciate how serious the situation is here and if we, together with Merriam [see below], cannot stop their talk about the earthquake problem. I for one am going to see what I can do about stopping the whole seismological game, and for the purpose of protecting the Institute.[18]

Robinson's threat was overtaken by the Long Beach earthquake, with Meltsner noting that:

> Robinson had two contradictory motives: he wanted to protect property values, but he also wanted to protect the California Institute of Technology. He wrote to Millikan because he was a well-known physicist, recipient of the Nobel Prize in 1923, but, more importantly, a very prominent member of the Institute and a member of the advisory committee in seismology of the Carnegie Institution of Washington. In the letter, he also brought in John C. Merriam [see above], the president of the Carnegie Institution, who was a supporter of seismological research. Since the Carnegie Institution, at the

13

time, provided most of the financial support for the operation of the seismological laboratory, and Wood and Buwalda were associated with the laboratory, Robinson probably thought that intimation of financial sanctions would cause one colleague to attempt to silence other colleagues.[19]

Structural Engineers Emerge as the Primary Seismic Safety Advocates

Structural engineering has its roots in the broader field of civil engineering. In fact, California's professional registration laws require a Civil Engineering license before the candidate can obtain one to practice structural engineering. In the decades after the 1906 San Francisco earthquake, the discipline of structural engineering came into its own largely because of the need for large buildings in the state's rapidly expanding urban areas. This new "craft" logically established its own professional associations: Southern California in 1929, Northern California in 1930, and a statewide organization in 1932.[20]

It seems likely that one strong impetus for these organizations was structural engineers' participation on the Committee of 100, the members of which were listed among the southern and northern sections of the American Society of Civil Engineers. The emergence of the structural engineer associations, and their continuing and largely voluntary work to prepare recommended earthquake-resistant design criteria and standards (commonly known as the "Blue Book") provided California with the professional and technical organizational capability to lead many seismic safety efforts.

The following two chapters (named for their authors) focus on the legislative politics involved with enacting The Safety of Design and Construction of Public Schools Buildings Act of 1933 ("Field Act") and California's Earthquake Protection Law ("Riley Act").

ENDNOTES

1. Reuben W. Binder, "Engineering Aspects of the 1933 Long Beach Earthquake," in *Proceedings of the Symposium on Earthquake and Blast Effects on Structures* (Oakland, CA: Earthquake Engineering Research Institute, 1952): 186.

2. C. R. Geschwind, *Earthquakes and Their Interpretation: The Campaign for Seismic Safety in California, 1906–1933* (Ann Arbor, MI: UMI Dissertation Services, 1996): 226.

3. Henry D. Dewell and Bailey Willis, "Earthquake Damage to Buildings," *Bulletin of the Seismological Society of America* 15, no. 4 (December 1925): 282–285.

4. "Report of the Engineering Committee on the Santa Barbara Earthquake," *Bulletin of the Seismological Society of America* 15, no. 4 (December 1925): 302–303.

5. F. M. Andrus, "Earthquake Design Requirements of the Uniform Building Code," in *Proceedings of the Symposium on Earthquake and Blast Effects on Structures* (Oakland, CA: Earthquake Engineering Research Institute, 1952): 314.

6. California Chamber of Commerce, *Building Code for California,* vii (1939).

7. *Engineering News Record* (April 6, 1933): 442.

8. *Engineering News Record* (June 22, 1933): 805.

9. Geschwind, 237.

10. W. M. Davis, "The Long Beach Earthquake," *The Geographical Review* XXIV 1, no. 1 (January 1934).

11. Charles Hillinger, "Dam Break Greatest Southland Disaster," *Los Angeles Times* (March 25, 1952).

12. Ibid.

13. Ibid.

14. *Sacramento Bee* (January 15, 1929): 19.

15. Ibid., 18.

16. *Sacramento Bee* (June 11, 1929): 4.

17. Ibid.

18. Henry M. Robinson, letter to Robert A. Millikan, April 8, 1927, as quoted in Arnold J. Meltsner, *The Communication of Scientific Information to the Wider Public: The Case of Seismology in California*, Minerva, XVII, No. 3 (Autumn 1979): 347.

19. Ibid.

20. Binder, *Earthquake and Blast Effects on Structures*: 187.

WAVE ONE, CHAPTER 2:
Protect Our Children: The Safety Of Design And Construction Of Public School Buildings Act Of 1933

The great and lasting good that came from the tragic and scandalous failures of school structures … was the passage of the Field Act by the state legislature…. I say "scandalous" failures because the strong shaking of the earthquake revealed shortcuts in construction practices, as well as design, in many school buildings. Experience in later earthquakes … has amply demonstrated the importance of the Field Act.[21]

California's first state earthquake safety law charged the Department of Public Works' Division of Architecture with regulating the design and construction of new public schools. The Safety of Design and Construction of Public School Buildings Act, commonly known as the "Field Act" for its author, Don C. Field, was triggered by the March 10, 1933, Long Beach earthquake. This law was one of two—the other being the "Riley Act" (see Chapter 3)—that created precedents for state seismic safety policy.

In fact, however, the question of the seismic safety of schools issue did not begin with the Long Beach earthquake, for in 1917, the United States Commissioner of Education issued a bulletin stating the following about the 1906 earthquake in San Francisco (which suffered so much):

> There is far more danger from possible earthquakes to poorly constructed buildings than from fire, for in the former case little time is offered to escape. Every school building in the city should be so constructed as to be more than reasonably safe from damage by earthquake.
>
> The people of San Francisco owe to the children of their city a large outlay for a large number of new school buildings thoroughly constructed against the danger of earthquake and also made entirely safe from fire. A city with the population and wealth of San Francisco, and with its comparatively small school population, should set the world a standard in the construction of safe and satisfactory school buildings.[22]

The report had negligible impact. The Long Beach earthquake would change that.

The Charter: Assembly Bill 2342 of March 23, 1933

Clearly, the structural engineers were intimately involved with the substance and politics of the Field Act:

> At another meeting which was held the following week (~March 21, 1933) much discussion concerned the problems of public schools and public buildings and possible legislation that would provide proper structural design of such buildings. J. B. Leonard of the Northern Association [of structural engineers] stated that the state architect, George B. McDougall, had been asked [by Assemblyman Field] to suggest legislation for design to withstand earthquake forces in public buildings, particularly in schools. Mr. McDougall in turn had telephoned him [Leonard] to ask whether there was unified opinion (likely among the structural engineers) as to what legislation was desired. *This exchange led ultimately to the passage of the Field Act, which assigned to the state architect the responsibility for the safe design and construction of public schools.*[23]

California Assemblyman Don C. Field, a Republican building contractor from Glendale whose district included earthquake-damaged communities—and who may have been in the area at the time of the March 10, 1933, earthquake because, according to one account he "witnessed the collapse of buildings"[24]—introduced AB 2342 on March 23, 1933. The Assembly's *Final History of the 1933 Session* (643) summarized the bill's key provisions:

> An act relating to the safety of design and construction of public school buildings, providing for regulation, inspection, and supervision of the construction, reconstruction, or alteration of or addition to public school buildings, and for the inspection of existing school buildings, defining the powers and duties of the State Division of Architecture in respect thereto, providing for the collection and distribution of fees, prescribing penalties for violation thereof, and declaring the urgency of the act, to take effect immediately.[25]

For the first time, California's state government was going to regulate the construction of public schools with the intent of avoiding the failures evident from the Long Beach earthquake, where:

> It is obvious that the time of the earthquake (5:55 p.m.) was fortunate in that loss of life was not great. If the time of the initial shock had been but a few hours earlier, the loss of life among school children would have been appalling.[26]

This observation about Long Beach was interestingly echoed by the Structural Engineers Association of Northern California (SEAONC) a year later as they reflected on the earlier June 29, 1925, Santa Barbara earthquake (which also occurred after school hours). SEAONC observed that "there would have been appalling injury and loss of life"[27] due to school damage if the earthquake would have occurred during school hours. The confirming Long Beach experience prompted action, and it is clear that California's structural engineers became strong policy advocates after the 1933 event:

Structural engineers were enjoying increasing prestige during this period (the 1930s). SEAONC sponsored legislative joint meetings in San Francisco in 1935, 1937, and later years. Many prominent members of the legislature and executive branch of the state were invited guests. The meetings were attended by representatives of SEAOC (Structural Engineers Association of California), ASCE (American Society of Civil Engineers), and other professional groups.[28]

AB 2342: The Politics

The formal processing of a piece of legislation is relatively easy to track, but the politics behind the legislation are much more difficult. Nonetheless, politics are the crux of any legislative story because it is how it "got done." In the case of the Field Act, we are very fortunate to have uncovered a fairly detailed insider story from an unpublished October 21, 1957, interview with a D. C. Willett, who in 1933 was an engineer serving as the Chief Assistant to the State Architect.

Known as "A Transcript of Conversation Between Mr. D. C. Willett and Mr. Frank Durkee," dated October 21, 1957, the material was compiled by Messrs. John F. Meehan and Donald K. Jephcott, both retired engineers from the state architect's office. The transcript appears as Appendix 1 to an unpublished report, "Task 4, The Review and Analysis of the Experience in Mitigating Earthquake Damage in California Public School Buildings." The authors prepared the document for Building Technology, Inc. in the summer of 1993.

Willett's account of the story behind the Field Act is fascinating, and the language and detail merit extensive quoting. It also reflects the very personal nature of most politics—but in particular the California politics of the time. In the exchanges below, D. C. Willett is "Mr. W" and "Mr. D" is interviewer Frank Durkee.

The first part of Willett's story focuses on the immediate reaction to the Long Beach disaster and the quest to "do something" at the state legislative level:

Mr. D: What did California have prior to that time [1933], if anything?

Mr. W: California had absolutely nothing in the form of a state code.

Mr. D: So then the people were aroused enough by this earthquake that had happened that our legislature decided to take some action to prevent such occurrences [building losses] in the future.

Mr. W: That's it. [I] think Don Field was the one that mainly was disturbed about what had happened—the buildings collapsing and everything—so he called Mr. McDougall [state architect] over to the Assembly Chamber and Mr. McDougall took me along—being the only engineer available, that is, in a supervisory capacity, so I was asked to go over there with him.

Mr. W: I think it was Saturday morning.... Anyway, we went over and met with the three of them, the senator and two assemblymen—Don Field being the main spokesman.[29]

Mr. W: The outcome was that I was ordered to contact the engineers in the state immediately.

Mr. D: The engineers. What do you mean?

Mr. W: The structural engineers, to develop the code requirements—a state code. Most of them were in Los Angeles. The different ones that had been working on code development were all down there.[30]

The schools were hit especially hard.... I told the ones [engineers] I could get on the phone that we wanted to get immediate action on a code. So that held me over the weekend. Monday at noon we were sitting at the desk eating lunch ... and the stories and pictures of the damaged schools were headlined.

Willett then shares that they "borrowed" some 1929 dam safety legislation as the template for what would become the Field Act. Willett also relates the truly startling fact that Field's original intention was not focused on schools. It was much broader, but political realities and problems of effective implementation ("enforcement") intervened:

> We were discussing the proceedings of what Assemblyman Field had asked us and how it could be worked out—and Fred [Fred Green, a senior engineer with the Division of Architecture] took the paper and he said, "You know what you ought to have—you ought to have a law governing school construction." "Well, gee," I said, "Fred, you've given me an idea." So I immediately went upstairs to the Division of Water Resources; they had an act [Dam Act of 1929] regulating the design and construction of dams ... to supervise the construction of dams all over the state.

> Mr. D: [So] now, Assemblyman Field, when he first presented this idea, wasn't directing it particularly at the schools then[?]

> Mr. W: No, no, he wanted a general code and I frankly told him he could not enforce a general code throughout the state.

> I went up and got the Dam Act and as soon as I finished lunch went over to the Assembly and got hold of Don Field. I said, "Don, I've got another idea. I don't know whether it's worth a darn or whether you want it or not, but here it is.... So I showed him the Dam Act; I showed him the pictures of the schools, and I said, "Now, listen, if you'll make a law to make school buildings safe, we can enforce it." I said, "We have the department that can put that over and make school buildings safe without any question, and there will be no trouble in the enforcement of it."

> Mr. D: Now, why would you expect that you could enforce such a law with respect to schools, but for general construction you were very doubtful that such a thing could be done?

> Mr. W: Well, of course, the schools were public money ... and it was the safety of the children, and the people would go for that where they wouldn't go spending their own money for

other safety measures. And not only that, the schools showed such tremendous defects in design and things that it was just ridiculous. So Don said, "This is just what we want." He grabbed it immediately.

I came back to see Mr. McDougall.... I went in and I said, "Chief, I've just been over talking to Don." I got that much out when the phone rang.

We cannot know to an absolute certainty, but it seems that Field then called Governor Rolph, who acted literally within minutes. Indeed, what would become the Field Act was crafted over a single weekend to take advantage of the political window of opportunity that Long Beach had opened. Even then it was not easy, at least when it reached the Senate and opposition began to form:

Governor Rolf [sic: Rolph] wanted Mr. McDougall to come to his office immediately. I said, "here's what he wants." I gave him the papers, explained to him what I'd told Field and the things Field wanted, so he grabbed them up and went to the governor's office.... He came back and said, "Now, we've go to get up a law. You work with the legislative counsel. Assemblyman Field would like to have this complete so he can present it to the legislature tomorrow. He says you can use the entire office. He says the office is at your disposal. Work up this law and get it in order, working through the legislative counsel and check with Don Field to see if it satisfies him." So we started in—practically the whole office outside the architectural group—working on different phases, and the girls typing it up. As soon as we got the thing roughed out we took it over to the legislative counsel. They jumped in.

Mr. D: Excuse me, what you did at the time then was—well, did you take this Dam Act and sort of revise it for your purposes?

Mr. W: That's right. We took the Dam Act as the foundation, and applied it to schools. The structural features and the regulations that would control schools were added and put that under architects and structural engineers.

Mr. W: He [the principal engineer of the Structural Section] was down there [Southern California] for several weeks.... It was more or less left to me but this was all done in a few days you understand, Saturday, Sunday. He didn't even know the school thing was cooking.... The thing that they were working on down there ... was to get a regular code—a general building code—because that was what he had been told to do. The development of a School Act was an afterthought which happened on Monday.

Mr. W: If you knew Mr. McDougall, he didn't want to be bothered with any detail of that kind [bill language].... He supervised it and I presented everything as we prepared it ... and he then told me to go to Don Field and do this and that, see. We worked on it the entire day and took it to the legislative counsel. They went through it and made some minor corrections and things.

Mr. D: That was Monday night[?]

Mr. W: That was Monday afternoon—we got it to them fast. It was really going. So after we got the thing assembled, roughed out, and typed, we presented it to Mr. Field. Mr. McDougall and I went over together and asked him how it looked. "Why you have thought of everything," he said. "This is just what I want." He said, "I'll tell what I want you to do. We won't have time to have this printed. I want you to have mimeographs made and put on every senator and assemblyman's desk so we can take it up the first thing tomorrow morning." He apologized for asking, but he said, "I think it's that important and you have to work tonight. Do whatever you can to get it mimeographed." So the next morning we had copies on everyone's desk.

Mr. W: So the oldest senator—I forget his name—was slow—and the bill didn't get put through in the Senate. The Assembly passed it unanimously.

Mr. D: Just as it was written?

Mr. W: Yes, just as it was written. It was passed with a unanimous vote in the Assembly. Then it was tied up. Mr. Leonard Starks, an architect in Sacramento, heard about it, and all thought we were slipping something over on the

24

architects and it was just ridiculous. So they managed to send telegrams and everything to block the action in the Senate. So the Senate recessed. The engineers, some of them came back from Los Angeles. Of course, you understand, the preparation of the bill was done without the knowledge of anybody so you can see why they would be suspicious. And apparently Don did some pretty shrewd work to put it through the legislature. He was going to get it through without getting any kicks.

Mr. W: Let's say they [architects and building industry representatives] blocked the act in the Senate. Well, it wasn't long until a group of engineers came up from Los Angeles. Earl Cope was president of the Structural Engineers Association of California at the time.

[H]e was a San Francisco engineer. Quite a few of them came up representing the engineers of the state and they wanted to know what the devil we were pulling on them. "Why, we haven't pulled anything," I said. "We used the Dam Act. If you can find anything wrong with it I would like to know." Mr. Cope read it and the statements, and he said, "Well, I'm for this." He said, "This is just what we should have and I'm for it, and I think the Association will back you to the limit on it." But the architects didn't like it.

Mr. D: Now why didn't they?

Mr. W: It forced them to hire structural engineers. There are very few architects that are qualified to handle the structural design of a building ... and they ... employed engineers to do this work. But, this bill, the way it was set up, more or less forced them to. And they know that and they didn't want that. The bill was coming up [for a hearing] and the opposition, the architects, were trying to build a little opposition, and along with this other opposition started in the Department of Education. The opposition did not especially come from the Department of Education as a Department, but came from the schoolhouse planning chief. The head of schoolhouse planning thought the bill should be under Education and not under the Division of Architecture.... But, knowing the educational set-up and knowing it to be dominated by educators, I told Don [Field] it would be just practically impossible if they wanted to enforce ... it under Education. It had to be free of any ties and

25

other things because we are going to have plenty of trouble enforcing it.

Willett then relates that the media, which at the time meant print media, were crucial in forcing the opposition to change their position:

> [A] Senate hearing was called. And prior to the Senate hearing ... the newspapers of the state got the architects together and they told them they would blast them in the headlines of the papers if they didn't go along with this bill; that work in the past was such that it needs supervision and that there was just no use of them opposing the adoption. So, as a result, the next morning, when the committee met, John Donovan, a prominent architect, gave one of the finest talks I have ever heard. Being an Irishman he could talk and had plenty of wit, an excellent speaker, and he gave one of the finest talks in support of the bill—which the day before he was bitterly against—and they saw that they had to come around.

Frank Durkee then questioned Willett about the role of public opinion in the aftermath of the Long Beach event. The answer was positive, but the supporting anecdote was particularly vivid, followed by a discussion of how any violation of the Field Act came to be a *felony*:

> Mr. D: At the time of the Long Beach earthquake, was there a great deal of public ... resentment against such poor construction? I mean was there a demand on the part of the public to do something or was this pretty much wholly from the legislative minds themselves, that they solve it?

> Mr. W: No, I think it was the public, as I had been told by an inspector on one of the [existing school] buildings, where one person was killed and others hurt, that the mob went out to lynch him. He had to get out of town.

> Mr. W: It was not his fault at all. I happened to examine the school that he inspected and he was following plans and specifications, but the plans and specifications weren't set up to withstand lateral forces. As a result there was a death in that school. But the thing that startled everybody, and this has been made public in statements by prominent educators,

is that there would have been at least 6,000 children killed had this happened during school time. That statement was made by the head of the school department after examining the buildings in Long Beach.... The main thing that had come up at the [Senate] hearing was that a violation of the code would have been a misdemeanor. That was in the Dam Act and we put it in this bill as being adequate. So a tall, dark-complexioned fellow ... with the Hearst Newspapers, the Los Angeles Examiner, demanded that it be made a felony ... it should be a felony for anyone to violate the Act.

I was sitting next to Don Field at the time.... He agreed to the change and I can say now that that was one of the greatest things that ever happened to the Field Act.

I think it was passed by a unanimous vote. Then it went through the [full] Senate. And of course, it had to go back to the Assembly for concurrence with the changes, as there had been quite a little rewriting—all to the advantage of the bill.

The way it finally came out, it was really a marvelous piece of legislation.

The engineers, everybody, afterwards got to feeling like ... Earl Cope, the president, when he said he was for it. Naturally, the engineers got their heads together and did anything they could to better it. They wouldn't let anybody lower the standards, yet they were perfectly willing to help in bettering the standards.

Mr. W: The Act itself had nothing to do with existing schools. It did give a [local school] board the right to ask for a structural examination and report on the buildings by the Division of Architecture. They pay the actual cost involved in such a report.

Later, however, Assemblyman Field found himself in political trouble in his home district—for reasons directly associated with the famous legislation. Field had a local, high-profile problem, felt that he had been betrayed, and was supremely angry:

About seven or eight months after the bill was adopted ... [Mr. McDougall] ... went ... at Field's request and saw Don. The Division had examined ... [the] big high school ... and we pronounced the building unsafe for use. This was in Glendale—Don Field's district. The building was [an architectural] monument. So they [local school officials] went out and hired a couple of civil engineers and their report on the building concluded that outside of some minor alterations the building was safe for use. That made Don Field pretty mad ... and he told him [McDougall] what he thought of the Division of Architecture.... [McDougall] said he would send Willett down and he would spend whatever time was necessary to find out whether the report was right or whether it was wrong and give him a definite statement. He [McDougall] told me to first contact Assemblyman Field, be sure that I knew everything he wanted and go through the building. If we had made a mistake, not to try to cover it up, but to come out and say so. So I went down there, and if I ever got a cussin', I did from Don Field. He was rabid. He told me that he had absolute confidence in me and in the Division and that's the reason he had put this bill through [the legislature]. He said we had apparently betrayed his confidence.

At two o'clock [on the day of Willett's arrival and after meeting with Assemblyman Field] I called him [Field] and said, "Don, every word in that report is the truth. There is nothing we could do otherwise. As far as I am concerned, the building is unsafe." He said, "Can you prove this to a layman?" I said, "I can prove it to anybody." He said, "Well, be at the school tomorrow at two o'clock. I want you to be able to show the people of Glendale why your report says this building is unsafe for use."

I might say at this time that the election was very close at hand. Don was apparently going to lose it because of this deal.

So ... at two o'clock he came and reported that there [were] three or four hundred people there. The whole side of the building was lined with people. They were fighting mad.

So, before we started the tour, Don got up in front on the steps of this monumental building and he told them what he had done, that Mr. McDougall had sent one of his best engineers

28

down here and ... he said, "I am turning you people over to Mr. Willett who says he can show you why this building is unsafe...." Then Don left [and Mr. Willett took people on a tour of the building, including the attic].

During part of the time while ... a bunch of newspapermen [were] there, a Colonel Evans, who was connected with the government on the loaning of money, looked over the situation and said—this was quoted, you'll see it in the papers—"Gentlemen, I wouldn't build a chicken house the way this school has been built." So, the result was that after the examination and after the report, people couldn't do too much to thank Don Field for what he had done on behalf of their children.

Naturally, after that, Don Field was absolutely sold on the Field Act and no one could touch it without asking us first. He asked many, many times when other bills or amendments came up, whether it would affect the Field Act or not. If we or Mr. McDougall said it was affected, he fought the thing to the end.

AB 2342: The Processing

D. C. Willett's story of the Field Act has to be complemented by a description of the more formal processing of the legislation, which has interesting aspects of its own.

On the day of AB 2342's introduction (the "first reading"), the Assembly unanimously consented to take the bill up on the floor, which eliminated referring the bill to committee, placing it on the "file" (to be heard at some future date), or sending it to the printer, any of which would have added time to the process. The Assembly also suspended other rules to accelerate it. AB 2342 was then read the second time, and the urgency clause was added. It was now "considered engrossed" (i.e., the original bill is compared with the printed version for accuracy purposes) whereupon it was read for the third time (on the Assembly floor). It passed unanimously and was sent to the Senate for its action. Some materials refer to a "committee" or a "commission" being appointed to hear AB 2342. No formal records of this organization were found, but it may have been an ad hoc mechanism to shortcut the process, because Willett's interview refers to meetings of senators and assemblymen.

On March 24, the Senate took up AB 2342. The bill took a more traditional route in this house, but it still moved relatively rapidly (the politics described by Willett were obviously behind the scenes). After its first reading, AB 2342 was referred to the Committee on Government Efficiency. Six days later, on March 30, the legislation was amended, and the amended version was sent to and returned from the printer. One amendment added the provision requiring the supervision of construction, and:

> Various other amendments ... were proposed at the hearing, among them one to compel inspectors on school buildings to make affidavit to the quality of materials used. Due consideration will be given to all proposed changes to strengthen the bill by the author, and enactment of the measure at an early date seems certain.[31]

On March 31 the committee sent the bill to the full Senate with a "do pass as amended" recommendation. On April 3, AB 2342 was made a special order of business for 11:30 a.m. on April 4, when it was referred to the Committee on Finance. On April 5, it was returned from the committee to the full Senate with a "do pass" recommendation. The urgency clause was read and adopted (a separate action), and AB 2342 passed the Senate unanimously to be returned to the Assembly for concurrence with the Senate's amendments.

The Assembly agreed with the Senate's language "without reference to committee," and AB 2342 was ordered to be enrolled (i.e., printed as a clean version by omitting symbols indicating amendments, and reviewed by the house of origin to see that the final text is in the form approved by both houses). Governor James Rolph Jr. received the "correctly enrolled" bill at 10:25 a.m. on April 7. He signed it into law on April 10—30 days after the earthquake—as Chapter 59 of the *Statutes of 1933*. AB 2342 took immediate effect because of its urgency clause, which stated:

> The series of earthquakes occurring in the southern portion of the state have caused great loss of life and damage to property. The public school buildings, constructed at public expense, were among the most seriously damaged buildings. Much of this loss and damage could have been avoided if the buildings and other structures had been properly constructed. The school buildings which will be erected, constructed, and reconstructed to replace the buildings damaged or destroyed by the earthquake should be so constructed as to resist, insofar as is possible, future earthquakes. These buildings will be erected, constructed, and reconstructed at once and, accordingly, it is necessary that this act go into immediate effect in order that the lives and property of the people will be protected.[32]

Early Implementation of the Field Act

Harry Bolin, a principal structural engineer for the State Division of Architecture's Los Angeles office, noted that "the Division of Architecture was given a duty quite different from any of those carried on before," and:

> Before enactment of the law, the Division of Architecture was concerned with the preparation of plans and specifications for preparation of estimates of, awarding contracts for, and supervising construction of state institutions, state buildings and such other buildings and structures as the state legislature designated. When the law became effective, the Division of Architecture entered a new field, viz., safeguarding occupants of public schools by checking plans of public school buildings as to adequacy and supervising construction thereof. In a sense, the Division of Architecture ... functions very much as a Department of Building and Safety for a municipality or a county, [but] the supervision of construction is more detailed and rigid.[33]

The Division of Architecture quickly had to establish implementing regulations. It accomplished this by adopting as "Appendix A" (later Title 21 of the California Administrative Code) the optional 1927 recommended appendix to the first edition of the Uniform Building Code—California Edition, which was similar to the ordinance that the City of Santa Barbara adopted, and was the beginning of code provisions for the seismic design of structures in the U.S. As noted earlier, the recommended code published in 1939 by the State Chamber of Commerce "was never adopted by any jurisdiction, but the provisions of the code were incorporated in the Uniform Building Code and in Appendix A."[34]

While the Field Act may not have been a "full employment act for engineers," it certainly did alleviate the impacts of the Great Depression on the profession by providing work and valuable experience for many engineers who emerged later, especially in the 1950s to 1970s, as leaders of California's earthquake engineering community. Numerous applications were received when the state announced that:

> An opportunity for structural engineers of high qualifications and experience to engage in the examination of important public buildings, including school houses, to determine their ability to withstand earthquake shocks, prepare structural plans, give consulting assistance to architects, structural engineers and contractors, check plans and supervise construction in accordance with standards set up by the state, is offered by examinations announced by the Division of Personnel and Organization, William Browning, chief, set for August 26, 1933. The examinations will be held in Los Angeles, San Francisco, Sacramento, and San Diego.[35]

A 1934 article by Clarence H. Kromer, a principal structural engineer for the California Division of Architecture, noted one early benefit of local compliance with Appendix A:

> It should be pointed out that buildings designed or reconstructed in accordance with Appendix "A" of the Division

of Architecture are in general subject to materially lower earthquake insurance rates than would be the case for similar buildings designed without any regard to bracing or earthquake resistance. This reduction in rates, depending on the probable resistance of the structure, is relatively large.[36]

The Field Act was not retroactive, nor did it provide the state with authority to condemn or close any existing public school buildings. However, the act did permit local school authorities, or a specified percentage (10%) of parents of enrolled students, to request school authorities to conduct seismic evaluations of their buildings. In little more than a year after the Field Act's implementation, Kromer's office had received over 333 applications for the review of plans for new schools, and over 1,000 applications had been filed for the voluntary examination of existing schools.[37]

The subject of school boards' liability apparently emerged quickly after the Field Act became law. This issue became more prominent later when it became apparent to the legislature that school boards were not strengthening or replacing deficient school buildings fast enough. Kromer summarized the situation as of 1934 when he wrote:

According to our understanding there is nothing in this act which places any additional responsibility on any school board other than that embodied in the existing statutes established prior to the enactment of this act. It does, however, provide the school board with a means of meeting its responsibility and even being relieved of it.

School boards throughout the state have been caused considerable concern by an opinion rendered by the attorney general to Mr. Vierling Kersey, superintendent of public instruction, under date of November 22, 1933. In this opinion, the question of liability or responsibility seems to hang on whether the district itself or its employees have been negligent in not taking proper precautions to provide safe buildings.

The school building act merely focuses attention on the matter of possible danger that might occur in the event of an

33

earthquake and provides a method whereby school boards may have authoritative information regarding the structural condition of their buildings.

Except for the fact that a warning has been sounded and possible danger to life and property recognized, it does not appear that it is any more necessary for school districts to go to the expense of making alterations or reconstructing their buildings than has heretofore been necessary nor is it any more necessary to close school buildings.

All that has been done is to attract attention and to emphasize responsibility of school boards, but the responsibility itself has not been increased.[38]

Senate Bill 797 was passed in 1935. It modified the original law regarding the liability and responsibility of school trustees. On January 6, 1936, a Senate Investigating Committee met on the University of California, Berkeley, campus to hear especially from representatives of the California School Trustees Association about "exactly what the limit of their [trustees'] responsibility is should anything happen to the children or to property due to an earthquake."[39] One of the committee's members was Senator J. C. Garrison, who later authored the 1939 Garrison Act, which specifically addressed the strengthening or replacement of pre-Field Act school buildings.

Later Looks at the Field Act

Nothing, however, is sacred or permanently guaranteed in the public policy arena. About 75 legislative bills have been introduced between 1934 and 2000 to amend or even abolish the Field Act. Some proposed amendments passed and others failed. The legislative proposals fall into several categories: (1) to require the retrofitting or replacement of pre-Field Act buildings; (2) to exempt various pre- or non-Field Act buildings for specific times or uses; (3) to raise funds to upgrade existing pre-Field Act schools; (4) to

34

eliminate or transfer to or share the state's responsibilities with local building safety agencies; (5) to "streamline" the Field Act's administrative processes; (6) to apply the act's principles to private schools; (7) to regulate portable or modular classrooms, especially their installation; (8) to require that school sites are evaluated before new schools are built or existing ones modified; (9) to include public "charter" schools within the Field Act's scope; (10) to conduct an inventory of early Field Act buildings (which, given current knowledge, might be hazardous); and (11) to prevent casualties from falling nonstructural building elements (e.g., lighting, ceiling panels, windows).

The next chapter focuses on the "Riley Act," which addressed buildings in general.

ENDNOTES

21. G. B. Oakeshott, *Volcanoes & Earthquakes: Geologic Violence* (New York City: McGraw Hill Book Co., 1976): 104.

22. U.S. Department of the Interior, Bureau of Education, "The Public School System of San Francisco, California: A Report to the San Francisco Board of Education of a Survey Made Under the Direction of the United Sates [sic] Commissioner of Education," *Bulletin* no. 46 (Washington, DC: GPO, 1917): 188–189.

23. Structural Engineers Association of California, *The Evolution of the Structural Engineers Association of California: Some Historical Notes, 1931–1981* (Sacramento, CA: Structural Engineers Association of California, 1981) (emphasis added).

24. A. E. Mann, *The Field Act and California Schools: A Report to the California Seismic Safety Commission* (SSC 79-02) (Sacramento, CA: Seismic Safety Commission, March 1979).

25. State of California, Assembly, *Final History of the 1933 Session,* (Sacramento, CA: State of California, Assembly, 1933): 643.

26. H. W. Bolin, "The Field Act of the State of California," in *Proceedings of the Symposium on Earthquake and Blast Effects on Structures* (Oakland, CA: Earthquake Engineering Research Institute, 1952): 309.

27. Structural Engineers Association of Northern California, *School Construction Under the Field Act* (March 1953): 1.

28. Structural Engineers Association of California, *The Evolution of the Structural Engineers Association of California: Some Historical Notes, 1931–1981* (Sacramento, CA: Structural Engineers Association of California, 1981): 6 (emphasis added). I recall hearing about these meetings being held until much later (late 1960s?) and being moved to Sacramento. A lack of legislative attendance and the fundraising expectations were given as the reasons for ending the practice. Some engineers saw the 1974 creation of the California Seismic Safety Commission as a continuation of this effort—author.

29. This may be the "Commission" referred to in some materials. I believe it probably was an ad hoc joint legislative committee charged to report directly back to each house within a few days—author.

30. This most likely refers to the committee mentioned in a paper presented before the Structural Engineers Association meeting in Santa Maria, October 16, 1936, by C.H. Kromer, in which he refers to the origin of Appendix A, now known as Title 24, California Administrative Code.

31. "State to Supervise School Building Plans," *Southwest Building and Contractor* (March 31, 1933): 13.

32. State of California, Chapter 59 of the Statutes of 1933 (1933).

33. H. W. Bolin, "The Field Act of the State of California," in *Proceedings of the Symposium on Earthquake and Blast Effects on Structures* (Oakland, CA: Earthquake Engineering Research Institute, 1952): 309–310.

34. E. G. Zacher, "History of Seismic Code Development in the United States," undated personal communication, probably given at a "Seismic Institute" held in the late 1970s or early 1980s: 1–2.

35. "Structural Engineers Wanted by State to Inspect Public Buildings," *Southwest Building and Contractor,* July 28, 1933: 17.

36. C. H. Kromer, "State Asked to Inspect More than One Thousand School Plants," *Southwest Builder and Contractor,* August 3, 1934: 12.

37. Ibid.

38. Ibid.

39. State of California, Senate, Notes of Meeting of Senate Investigating Committee, Safety of Design and Construction of Public School Buildings, Chapter 59, Statutes of 1933 (unpublished), Sacramento, CA, January 6, 1936: 1.

WAVE ONE, CHAPTER 3:
Design New Buildings For Earthquake Forces: The "Riley Act"

"A building is not a pile of elements but a unit, and it is wholly reasonable and absolutely necessary to consider it as such in any region known to be subject to earthquakes."[40]

California's Earthquake Protection Law (the "Riley Act") was one of two first-ever, state-level seismic safety legislative policy initiatives. The Riley Act addressed the earthquake-resistant construction of new buildings, and its companion initiative, the "Field Act," put state government squarely into regulating the construction of new public schools. These initiatives forever changed—in fact, created—the precedents for state seismic safety policy.

Both laws resulted from the March 10, 1933, Long Beach earthquake, leading noted geographer W. M. Davis to write in 1934:

> The ... earthquake ... will be less remembered by reason of its contributions to seismology—for as a crustal tremor there was nothing especially remarkable about it—than it will be for having broken down the "hush-hush" policy that has hitherto been followed by the commercial organizations of the cities of Southern California.[41]

The Charter: Assembly Bill 2391 of April 25, 1933

Harry Riley, a lithographer by trade, was representing Long Beach in the Assembly on March 10, 1933. The legislature was in session, and Mr. Riley was excused to return to his district because of the earthquake. Three days after the earthquake, Assemblyman Riley returned to Sacramento "to request a loan from the state of California of $500,000 to feed the homeless."[42] Although the amount was reduced to an initial $50,000 (more came later), the emergency bill was sponsored jointly by Mr. Riley and Assembly Speaker Walter J. Little, with the *Los Angeles Times* noting:

> Discarding committee meetings and other routine affairs, the Assembly rushed the legislation through ... and the Senate ... held itself in readiness for the measure long after the usual hour of adjournment. When it appeared, Lieut.-Gov. Merriam sped it through the Senate and in a matter of minutes it was on its way to the governor's desk for signature. He cancelled other engagements to wait for it."[43]

Serving as the triggering event, the Long Beach earthquake changed the normal political agenda and created a climate for further action. A *Los Angeles Times* March 13, 1933, story, "Gears Greased by Legislature," noted:

> Whatever legislative relief is needed to ease conditions in the Southern California earthquake zone will be speeded through the lower house by Speaker Little, Chairman [Lawrence] Cobb of the Ways and Means Committee and the entire Southern California delegation, with all other Assemblymen in sympathetic accord. Then it will be dumped into the Senate, where fast action is assured. The governor will complete the cycle by signing any measure in this respect as quickly as it reaches his desk.[44]

Assemblyman Harry Riley, like his colleague, Assemblyman Don C. Field, a building contractor from Glendale (who represented damaged portions of the City of Los Angeles and nearby communities), was to become an agent of state policy change when Mr. Riley introduced Assembly Bill 2391 on April 23, 1933. Mr. Field had introduced a related measure, Assembly Bill 2342, a month earlier (March 23, 1933). Mr. Field's legislation would require the state of California to set standards and approve plans for and supervise the construction of new public school buildings. Together, the Field and Riley acts symbolized a larger movement supported by an emerging and technically oriented "earthquake community" directed at improving codes, standards, and practices governing the earthquake-resistant design and construction of California's new buildings.

In a 1934 retrospective article ("Earthquake-Conscious?") that reflected on the emerging seismic safety trend, the *Engineering News Record* (ENR) noted:

> The statewide movement toward designing against seismic risk ... is unlike anything that has occurred heretofore. It is an extremely healthy and commendable trend. Previously private owners [probably only a few—author] have seen the advantage of designing against a reasonable amount of lateral force. Now all building in California cities must take this factor into account, and special stress is placed on schools with attendant responsibility on public officials. Apparently, an "earthquake-conscious" viewpoint has come to stay.[45]

AB 2391's Path to Passage

While its intent probably was quickly forgotten, Mr. Riley's AB 2391 was introduced on April 25 "with the purpose of *controlling reconstruction* following the earthquake of March 10 [1933]. Opposition was mainly in regard to the provisions of the bill and not to its intent."[46]

Building code requirements existed in the City of Long Beach before the 1933 earthquake, but they may not have contained earthquake-resistant design requirements. The following report of a discussion between C. C. Wailes Jr., chief building inspector at the time of the earthquake, and Edward M. O'Connor, a successor to Wailes (and a colleague of the author's), shows that the city attorney effectively prevented the city's building department from requiring owners to repair (reconstruct?) damaged buildings to resist earthquakes:

> He (Wailes) advised me that he spoke to the city attorney ... and asked him if he could request that the owners of damaged buildings repair such buildings in an earthquake-resistant manner. The then city attorney advised him that he could not request that a new building be built in an earthquake-resistant manner until the Long Beach regulations were amended to require new buildings to be designed and built in an earthquake-resistant manner. The city attorney advised him that the only thing that he could require for the repair of existing buildings was that they be repaired in a manner so that they were no worse than they were prior to the 1933 earthquake. The building regulations of the City of Long Beach were amended in January 1934, to require that new buildings be designed and built in an earthquake-resistant manner.[47]

The Riley Act followed the Field Act through the legislative process. Soon after AB 2391 was introduced, it was referred first to the Assembly's Committee on Public Health and Quarantine (on which Assemblyman Field sat), and then on the same day it went to the Committee on Revenue and Taxation (chaired by Mr. Riley). On May 2, 1933, AB 2391 went to the Assembly floor with a "do pass as amended" recommendation. The urgency clause was added on May 5, and it passed the full Assembly on that day.

One week later (May 12), AB 2391 was heard by the Senate's Committee on Municipal Corporations, which gave the bill a "do pass" recommendation, whereupon AB 2391 went to and was passed by the full Senate on the same day. AB 2391 also was returned for final concurrence on May 12 to the Assembly, because that was the bill's house of origin. The Assembly concurred in the bill's final language, and after the enrollment of AB 2391 on May 15, it was presented to Republican Governor James Rolph Jr. at 11:30 a.m. for his signature.

Governor Rolph signed AB 2391 into law as an urgency statute on May 26, 1933, where it became Chapter 601 of the "Final Calendar of Legislative Business, 50th Session, 1933." The urgency clause was the same as the Field Act's, and it sounds very similar to those used for post-earthquake seismic safety legislation today:

> The series of earthquakes occurring in the southern portion of the state has caused great loss of life and damage to property. Much of this loss and damage could have been avoided if the buildings and other structures had been properly constructed. The buildings which will be constructed and reconstructed to replace the buildings damaged or destroyed by earthquake should be so constructed as to resist, in so far as possible, future earthquakes. These buildings will be constructed and reconstructed at once and accordingly it is necessary that this act go into immediate effect in order that these buildings be so constructed that the lives and property of the people will be safeguarded.[48]

While there was some controversy about several of the bill's provisions, one article noted:

> After some slight changes in the original measure it developed that, in the rush of closing legislative action, any attempt to amend the bill further would kill it and, in the interest of having the measure as it stood or none, opposition was withdrawn.[49]

While awaiting Governor Rolph's signature, an article about Mr. Riley's AB 2391 reflected how many organizations had come together to endorse the pending law: evidence of a strong consensus and the presence of an "advocacy coalition."[50] The cited organizations included the Structural Engineers' Association of Southern California, its Northern California counterpart, the statewide Structural Engineers' Association, the chapters of the American Institute of Architects of the Northern and Southern (California) Districts, and the California State Chamber of Commerce. Interestingly, the article also noted "The proposed act is not intended to limit the powers of municipalities or counties to establish higher standards, its real object being to insure better construction in all communities throughout the state."[51]

Legislative Intent

The intent of the relatively short and simple Riley Act was "to regulate the construction of (new) buildings in the state of California, in respect to horizontal forces, providing penalties for the violation thereof and providing that this act became effective immediately."[52] The exceptions included buildings not intended primarily for human occupancy, dwellings for not more than two families, and buildings under construction at the time of enactment.

Where did the original law's proposed language come from? To answer this question we must return to the 1925 Santa Barbara earthquake and to one of its legacies: the adoption by the cities of Santa Barbara and Palo Alto of new building codes that for the first time contained earthquake design requirements. We also must recognize the work of a "Committee of 100" working under the auspices of the State Chamber of Commerce on a "California Building Code"; findings of a Los Angeles County Coroner's jury that convened soon after the earthquake; the roles played by members of the emerging "earthquake community," especially those belonging to the new Structural Engineers Association of Southern California (SEAOSC, 1929), Northern California (SEAONC, 1930), and the statewide association (SEAOC, 1932); and from the membership, processes, and the 1933 final report of the Joint Technical Committee on Earthquake Protection, chaired by Robert A. Millikan, president of the California Institute of Technology (Caltech).

Although direct evidence is a little obscure, it strongly appears that Assemblyman Riley depended principally on members of the SEAOC, including its southern and northern chapters, to provide the ideas, technical language, and continuing support he needed for the legislation.

Pre-1933 Activities

Reuben W. Binder, then a practicing engineer with Bethlehem Pacific Coast Steel Corporation, provided a useful summary in 1952 of pre-1933 earthquake engineering knowledge and practice, applicable portions of which follow:

> The year 1925, in my opinion, marks the real beginning of earthquake engineering studies and research in the United States. From that year on, the interest and accomplishments in connection with earthquake engineering studies advanced at an accelerated pace.
>
> The Santa Barbara earthquake of 1925 was the focal point for again arousing the interest of many practicing structural engineers in earthquake phenomena as affecting design.
>
> Following the Santa Barbara shock, earthquake-resistant requirements began to be incorporated into building codes. The first code to add earthquake provisions was that of Santa Barbara in 1925. The Palo Alto code followed with similar regulations.
>
> The first edition of the Uniform Building Code of the Pacific Coast Building Officials Conference, published in 1927, contained in the appendix a chapter on earthquake provisions for optional use.
>
> In 1928, the California State Chamber of Commerce sponsored the preparation of a Building Code for California. This project stimulated the wide-spread and active interest among engineers and architects in the subject of earthquake-resistant design.
>
> Interest in engineering problems in connection with earthquake studies influenced the formation in 1929 of the Structural

Engineers Association of Southern California. In 1930, the Structural Engineers Association of Northern California was organized, and in 1932 the Structural Engineers Association of California came into existence.[53]

The State Chamber's Building Code for California

Although the State Chamber of Commerce initiated work on a recommended building code after the 1925 Santa Barbara earthquake and before the 1933 Long Beach earthquake, it was not completed until 1939. This work-in-process probably was important to the Riley Act's development, and the more than 100 volunteer "highly qualified technical men" from Northern and Southern California (labeled later as the "Committee of 100") also benefited from the knowledge gained from the Long Beach earthquake.

Charles F. Richter, the noted Caltech seismologist, summarized the earthquake insurance "crisis" following the Santa Barbara earthquake when he noted:

> Prior to that, earthquake insurance had been written very extensively in Southern California with little regard to the actuarial soundness, and that event was a great shock with some companies suffering losses in claims that were disproportionate for a comparatively moderate event.[54]

The State Chamber's document further reflects the concerns of California's business interests:

> The earthquake of 1925 which centered at Santa Barbara, and caused heavy building damage there, resulted in a sudden and widespread demand for earthquake insurance. This in turn was followed immediately by an increase in earthquake insurance rates and in the amount of such insurance required by the State Corporation Commissioner before he would approve bond issues on certain types of buildings.
>
> The result of this twofold handicap was a sharp recession of building in the state, and in the emergency, the aid of the

45

State Chamber was urgently sought by business interests generally. The Chamber undertook an investigation [that] at once resulted in:

1.   Reduction of earthquake coverage requirements by the State Corporation Commission.
2.   Reclassification of buildings by the Board of Fire Underwriters and a readjustment of earthquake insurance rates.

But it soon became evident that the fundamental need was for a statewide building code which would require adequate standards of building construction and a reasonable resistance of such construction to earth movements, and to this task the State Chamber directed its attention.[55]

Upon the recommended code's completion, the State Chamber thanked its committee members and the business and industrial interests that financially supported the effort, finding "that the group best equipped to promulgate [seek adoption—author] the Building Code for California is the Pacific Coast Building Officials Conference."[56]

Research political scientist and longtime California Seismic Safety Commissioner Stanley Scott observed that the Chamber's Building Code for California was "never adopted as a code anywhere, (but it) was nevertheless considered a good document for the time." It became the Field Act's initial regulations ("Appendix A"), but it did "reach beyond the legally specified target of public schools ... causing an appreciable upgrading of design and construction practice that carried over into the private sector and other types of structures."[57]

Los Angeles County's Coroner's Jury in Earthquake Inquisition

Officially convened by Frank A. Nance, County Coroner, "In the Matter of the Inquisition Upon the Bodies of Walter Benson De Buxton and Tony Gugliemi, Deceased," the

Coroner's Jury's 3,500-word verdict on March 28, 1933, primarily addressed Los Angeles County's needs for better building standards.

The jury was composed of nine "experts in building construction but also men qualified to render adequate judgment on every phase of the earthquake disaster." Coroner Nance's jury, "following the policy adopted at the time of the St. Francis Dam disaster" (1928, which led to the state's Dam Safety Act—author), was charged with making "a detailed report which will not only explain the factors responsible for ruin and injury attendant on the quake, but may serve as a guide to preventive measures that will lessen the possibilities of bodily injury and property damage."[58]

One of the jury's final recommendations was that Los Angeles County adopt the Pacific Coast Building Officials' (later the International Conference of Building Officials' (ICBO) recommended code, work on which began soon after the Santa Barbara earthquake, meant that a common technical basis could be used for the Riley and Field Acts. The *Engineering News Record* noted, "the result will have greater weight because the public will base its opinions on the findings of this official body and thus establish a starting point for much of the coming permanent reconstruction" (see the Riley Act's original intent: to control reconstruction—author).[59]

Indirectly, the verdict also provided ideas for state legislation:

None of the [local] ordinances in any portion of the damaged area adequately provided protection against horizontal or other stresses produced by earthquake oscillations. Stability of construction was often sacrificed for architectural effects or for purposes of utility or convenience, or with an eye to economy.

Damage was the greatest in cases where emphasis was placed upon quantity rather than quality, effect rather than stability, or fostered by the thought of constructing only to skimpily meet minimum regulations.

Materials normally good under certain conditions were often used in places unsuited to their nature, or the purpose which they were intended to serve was nullified on account of poor workmanship, or other inferior materials accompanying them, or because of improper installation.[60]

4 – That the State Contractor's License Law be strengthened so as to require still greater ability and integrity on the part of contractors.[61]

California's Architects Rally

Further support for state legislation was reflected in a set of resolutions "petitioning the legislature to pass an enabling act making it mandatory on every district (local government) in the state to make its building regulations more rigid." The subject resolutions resulted from an April 4, 1933, joint session of the Southern California Chapter of the American Institute of Architects and the State Association of State Architects. John C. Austin, chairman, in his opening remarks said:

It has taken an earthquake to bring us together. I believe we shall do what Western men have always done in an emergency—band together to bring order out of chaos. Lessons learned from other earthquakes and talk of regulating ordinances were soon forgotten. We must now provide laws that will create the best type of building throughout the state and work for better ordinances.[62]

A letter from Dr. Robert A. Millikan, president of the California Institute of Technology (Caltech), was read to the group. It noted that he had accepted the chairmanship of a "group of organizations (the Joint Technical Committee on Earthquake Protection—see below) correlating all activities on reaching conclusions as to procedure in securing adequate state legislation, safeguarding against building failures in future emergencies that may arise."[63]

The Joint Technical Committee on Earthquake Protection

The Los Angeles Chamber of Commerce began forming the Joint Technical Committee on Earthquake Protection within about two weeks after the Long Beach earthquake. As noted, Robert A. Millikan, president of Caltech and a distinguished scientist, social activist, and supporter of earthquake research, was recruited to chair the committee. Typical of the letters he received was one from Lynn Atkinson, then president of the Southern California Chapter of the Associated General Contractors:

> There are numerous committees and groups interested in discussing the results of the earthquake, and future precautions. These various organizations and groups lack coordination and leadership, and much of their well-intentioned and competent efforts will be wasted unless the endeavor is organized.
>
> It is has been suggested that a joint committee composed of representatives of the various interested groups be organized ... (but) the appointment of any prominent representative within the industry might prejudice the activity of such a committee due to jealousies or imagined prejudices, and for this reason it has also been suggested that someone in a neutral position should head up such a committee.
>
> Your name has been suggested and ... many of the more prominent organizations and groups have endorsed this suggestion.... I am therefore taking this opportunity to urge upon you the acceptance of such chairmanship.
>
> Unless such a committee headed by a man such as yourself ... it is quite possible that political and publicity elements may dominate what is largely a technical situation that should be handled by technicians thoughtfully. Hasty and incompetent legislation may also result.[64]

Dr. Millikan accepted, and referring to the joint meeting of the architects and engineers discussed above, "Several

49

hundred architects and engineers were at the meeting and this [Millikan's acceptance] letter brought much applause.... Surely we could not have found a better man[;] here at last the brick lobbyist will find his match."[65]

The 19 members of the Joint Technical Committee proceeded quickly, completing their final 13-page report ("Earthquake Hazard and Earthquake Protection") in June 1933. While it did not call directly for state legislation, the report was widely acclaimed and provided a latent "policy agenda." It talked about regional earthquake history and seismicity ("this region is seismically active"); various effects of the Long Beach event; standards for the design of new structures; strategies for strengthening existing buildings (the retroactive issue— author); prevention of conflagrations (referring to the 1906 San Francisco and 1923 Tokyo earthquakes); utilities; and disaster planning.

The report's cover letter said:

> We believe that a comprehensive statement of the true facts, based upon scientific findings and building construction truths, is most timely and instructive and certainly should be considered carefully by all public bodies, institutions, and individuals who are in any way responsible for the safety of lives and property in this area.

One of the report's truths related directly to the intent of the Riley Act was:

> Only general mention need be made of an all too prevalent class of construction in this region; namely, the stores and apartment houses which were built as cheaply as the inadequate building codes would permit by those who were interested only in speculative profits.... Hundreds of such examples of unsound building construction were wrecked, and as a class they account for the large part of the toll of property damage.[66]

Further Insights into the Riley Act: The D.C. Willett Interview

Retired Supervisory Engineer D.C. Willett of the Office of the State Architect in Sacramento was interviewed in 1957 about his activities associated especially with the Field Act. However, portions of the unpublished interview remind us that Assemblyman Field "was the ringleader in the group and he didn't know what he wanted," but he originally considered introducing legislation (for a "state code") that would have prescribed lateral force (earthquake) standards for all buildings in California."

Mr. Willett, being the only supervisory engineer available because others already were in Los Angeles to help with building inspections, accompanied George B. McDougall, then the state architect and a governor's appointee, to a meeting at the Capitol with Assemblyman Field and another unnamed assemblyman and a senator. During the meeting D. C. Willett noted, "I frankly told him (Field) that he could not enforce a general code throughout the state."

These discussions led Assemblyman Field to focus his efforts on public schools only because state law mandated that children attend schools and that state money was used to build many school buildings, providing a strong rationale for direct state intervention in school building design and construction.

Mr. Willett's supervisor, the principal engineer who headed the Structural Section, was one of those in Southern California after the Long Beach earthquake. Willett noted that "(the principal) was working ... to get a regular code—a general building code—because that was what we had been told to do."[67]

Now that schools were going to be addressed at the state level by Assemblyman Field, this left the issue of general building standards open in the legislature, and if statewide policy was going to be established it would have to be addressed by someone else. Who better than Assemblyman Harry Riley from Long Beach?

D. C. Willett noted further that after the Field Act was signed into law, Assemblyman Field:

> ... lost interest in a general code (but) the legislators felt that there should be some kind of legislation, so they set up restrictions in the Riley Act—very, very small ones. I think it was the structural engineers who more or less sponsored it. They were demanding that something be done. So they set up the Riley Act which, at first, took care of a lateral force of 2% which was all they could get. But it should have been 10%.

> That was all they [the structural engineer sponsors] could put over. Nevertheless, they did get one thing in that they wanted—a 20 lb. wind, which was more or less equivalent to around 10% (lateral), but nobody realized the wind governed. It got no opposition. I was talking with Earl Cope, one of the (structural engineer) leaders in the thing. I said, "Earl, do you know what this wind means?" He said, "Keep quiet." So, the 20 lb. wind was the governing feature of the Riley Act. But it was not enforceable. In other words, the Riley Act for years and years didn't amount to anything. There were amendments after amendments. The 20 lb. wind was finally lowered.[68]

Early Implementation of the Riley Act

For the first time, the State of California set standards for the earthquake-resistant design of all new buildings except certain types of dwellings and farm buildings. The provisions of the new California Earthquake Protection Law (Riley Act) became effective immediately, with enforcement left to local governments.

However, noting that "the present act is a promising start," concerns [probably those set aside to get the law passed—

52

author] about the Riley Act's implementation quickly emerged. Five days after the governor signed AB 2391, the *ENR* noted that "while the law is highly commendable":

Its practical aspects are more doubtful. Drawn in haste to meet the emergency, the act promises to involve serious administrative problems and to affect present practice and building costs extensively. Amendment to make it less burdensome is certain to come as soon as the state's requirements [implementing regulations—author] can be studied more carefully.

The most glaring defect of the bill is the uniformity with which it applies to the entire state, though it is well known that there is wide variance in the susceptibility of different regions to seismic disturbance. Hardly less objectionable is the lack of distinction between foundation conditions, the requirements being alike for solid rock land and soft fill. The indefiniteness of its application to practical design, especially in respect to stress intensities, will lead to varying interpretation by the municipal building departments that must administer the act, and designers will therefore be confronted with a complexity of rules and regulations.[69]

Interpretation

The Long Beach earthquake opened the window of opportunity for enacting California's first state-level seismic safety legislation. Although seemingly intended to govern reconstruction in the damaged area, Assemblyman Harry Riley's sponsorship of AB 2391, while weak in some respects, built upon some local initiatives, was generally consistent with several studies, appears to have involved a broadly based and technically oriented advocacy coalition, and at least provided a statewide foundation for including lateral force requirements in building codes.

Earthquake engineering knowledge, experience, and research has continued in the subsequent decades, and the earthquake standards contained in locally adopted building codes (the Uniform Building Code ["UBC"] in California) have effectively superseded the Riley Act's provisions. Central to the earthquake community's abilities to seize the 1971 San Fernando earthquake legislative window of opportunity was the building of their professional and technical capabilities in the intervening decades, during which, incidentally, no major state-level seismic safety legislative initiatives emerged.

Unlike the Field Act, AB 2391 belongs to that category of programs that are enacted by the state but are administered by local government. While the Riley Act has been amended, no organizational focus for its administration exists in state government. No records exist about how many local governments relied on the state for the enforcement of the Riley Act, but although evidence is scarce, it is clear that at the time of the Long Beach earthquake very few local governments had adopted building regulations and even fewer had effective enforcement programs managed by "building safety departments." An *ENR* article noted that "Enforcement is placed under municipal buildings departments and where these do not exist the state division of architecture is to control."[70] The act itself said:

> If no such (local) ordinance is in effect at the place at which the work is done then the allowable working stresses shall be those specified by the Division of Architecture of the State Department of Public Works, which is hereby fully authorized and empowered to specify such allowable working stresses for the purpose of this act as to any city, city and county, or county in which no such ordinance is in effect.

The state now mandated local governments to enforce a minimal "earthquake code," and we can also reasonably assume that the Riley Act contributed to the establishment of local building regulatory agencies during the next few decades, which today remain California's primary code adopting and enforcing agencies. Although its provisions remain in the statutes (Health and Safety Code Sections

19100–19170), California's local governments now routinely meet or exceed the Riley Act's requirements. Nevertheless, it seems clear that without the act it would have taken much longer, perhaps never, to set a statewide minimum design standard for all new buildings.

Less than a year and a half after the earthquake another *ENR* article stated that several Southern California cities (Santa Monica, Los Angeles, Beverly Hills, Long Beach, Pasadena) had adopted lateral force provisions in their building codes containing greater requirements ("seismic factors") than the Riley Act's minimal 2 percent.[71]

The act's original lateral force design requirements of 2 percent of the total vertical design load were changed administratively in 1953. The change required use of a 3 percent lateral force factor for buildings less than 40 feet tall and a 2 percent factor for those over 40 feet tall. In 1965 the Riley Act's lateral force requirements again were modified to simply require that the applicable parts of California's Administrative Code (containing agency regulations) become the Riley Act's governing lateral force requirements, which "...were essentially the same as those given in the 1961 Uniform Building Code and which *followed the recommendations of the Structural Engineers Association of California.*"[72]

It is clear also that the new and politically active "earthquake community" was the real sponsor, establishing a tradition of political action that continues. Although Assembly Bill 2210 of 1971 tried, the Riley Act was never made retroactive so it would have applied to pre-1933 buildings. Other strategies and programs at later times had to address this difficult issue.

Notably because the Joint Committee on Seismic Safety's 70 volunteers mobilized as an advocacy coalition immediately following the February 9, 1971, San Fernando earthquake, Wave Two, Chapter 4 discusses a simple law that required

California's local governments to address their earthquake threat by requiring a Seismic Safety Element be added to their general land use plans.

ENDNOTES

40. "Reasonable Preparedness," *Engineering News Record,* April 13, 1933: 479.

41. W. M. Davis, "The Long Beach Earthquake," *The Geographical Review* XXIV, no. 1 (January 1934): 1.

42. *Los Angeles Times,* March 13, 1933: 2.

43. *Los Angeles Times,* March 14, 1933: 1.

44. *Los Angeles Times,* March 13, 1933: 8.

45. Engineering News Record, July 5, 1934, 21.

46. "New California Building Law Makes Lateral Force Mandatory," *Engineering News Record,* June 1, 1933: 722 (emphasis added).

47. Edward O'Connor, "Pre-Field Act Buildings in the City of Long Beach" (Proceedings, 39th Annual Convention, Structural Engineers Association of California, October 15–17, 1969): 20.

48. *Southwest Building and Contractor,* May 26, 1933: 8.

49. *Engineering News Record,* June 1, 1933: 722.

50. An advocacy coalition consists of actors from a variety of public and private institutions at all levels of government who share a set of basic beliefs ... and who seek to manipulate the rules, budgets, and personnel of governmental institutions in order to achieve these goals over time. (Sabatier & Jenkins-Smith, 1993, as quoted in Olson, Olson and Gawronski, *Some Buildings Just Can't Dance: Politics, Life Safety, and Disaster,* 1999: 15.)

51. *Southwest Building and Contractor,* May 26, 1933: 8.

52. State of California, *Final Calendar of Legislative Business,* 1933: 657.

53. R. W. Binder, "Engineering Aspects of the Long Beach Earthquake," (Proceedings on Earthquake and Blast Effects on Structures, Earthquake Engineering Research Institute, 1952): 186–187.

54. "Charles F. Richter – How It Was," oral history, *Engineering & Science,* March 1982: 26.

55. California State Chamber of Commerce, *Building Code for California,* 1939, vii.

56. Ibid., viii.

57. Stanley Scott, "Earthquake Engineering: Conditions Influencing the Practice," (draft working paper, Institute of Governmental Studies, University of California Berkeley, May 1990).

58. *Los Angeles Times,* March 14, 1933: 3.

59. *Engineering News Record,* April 13, 1933: 478.

60. Frank A. Nance, County Coroner, "In the Matter of the Inquisition Upon the Bodies of Walter Benson De Buxton and Tony Gugliemi, Deceased" (findings of the Coroner's Jury, March 28, 1933, 7: lines 5–19 and Recommendation).

61. Ibid., 10: lines 3–5.

62. *Southwest Building and Contractor,* April 7, 1933: 12.

63. Ibid., this same issue also noted that Governor Rolph had signed the Field Act into state law.

64. Lynn Atkinson, letter to Robert A. Millikan, March 23, 1933.

65. Arthur R. Hutchason, letter to Bernhard Hoffman, April 5, 1933.

66. Joint Technical Committee on Earthquake Protection, "Earthquake Hazard and Earthquake Protection," June 1933: 5.

67. The principal may have been working with the architects and engineers—perhaps the Committee of 100—to prepare a recommended general building code, and it so happened that a legislative opportunity to set minimum statewide standards presented itself—author.

68. Ibid., 22–23.

69. *Engineering News Record,* June 1, 1933: 721.

70. *Engineering News Record,* "New California Building Law Makes Lateral Force Design Mandatory," June 1, 1933: 722.

71. *Engineering News Record,* "California Makes Progress Against Earthquake Hazard," July 5, 1934: 15. These communities might have used the Pacific Coast Building Officials' suggested code rather than enforce the Riley Act, partially because if they did nothing the state might intervene—author.

72. J. F. Meehan, "History of Earthquake Codes in California," Nov. 8, 1972 (unpublished background paper for the Joint Committee on Seismic Safety's Final Report, 1974: 195, emphasis added).

Part II. Wave Two

WAVE TWO, CHAPTER 4:
Planners—Pay Attention To Earthquake And Geological Hazards!

"The San Fernando earthquake has given city planners good reason to re-evaluate the land use planning process with respect to seismic safety."[73]

The Charter: Senate Bill 351 of February 19, 1971

Senator Alquist, chairman of the Joint Committee on Seismic Safety, introduced legislation to require seismic safety be considered by local planners (Senate Bill 351) and the original hospital seismic safety act (Senate Bill 352) on February 19, 1971, just ten days after the San Fernando earthquake. When putting the two bills "across the desk" to start the legislative process, he noted, "I am convinced that the loss of life and property in the Los Angeles earthquake could have been reduced had these requirements been in effect."[74]

SB 351 cleared both legislative houses on its first attempt, and within five months of its introduction SB 351 was signed into law by Republican Governor Ronald Reagan on June 17, 1971, becoming effective 61 days after adjournment as Chapter 150 of the Statutes of 1971 (Government Code Section 65302). SB 351 was one of the earliest pieces of legislation introduced after the 1971 San Fernando earthquake, and it was the Joint Committee on Seismic Safety's (JCSS) first substantive legislative success.

George G. Mader, a planner, member (and later chairman) of the JCSS' Advisory Group on Land Use Planning (AG/LUP), and consultant to the JCSS on the San Fernando earthquake study, captured the intent of recently enacted SB 351 when discussing it at a 1971 League of California Cities meeting:

> Members of the Advisory Group and Land Use Planning realized there is much to be learned and developed with respect to controlling urban growth in relation to seismic hazards. Nonetheless, the consensus was that rather than wait until a perfected methodology is available, it would be more prudent to add a very broad provision to the State Planning Law which would require cities and counties to take seismic hazards into account in their planning programs. The law is therefore worded very generally yet comprehensively with respect to the types of seismic hazards that should be considered.[75]

It was a simple law, comparable in scope to other prescribed elements of city and county general land use plans. This one required local governments to prepare: "(f) A Seismic Safety Element (SSE) consisting of an identification and appraisal of seismic hazards such as susceptibility to surface ruptures from faulting, to ground shaking, to ground failures, or to effects of seismically induced waves such as tsunamis and seiches."

The 1971 legislative requirement for an SSE was reinforced later (1975) by requiring a separate "Safety Element." The newer law required:

> A safety element for the protection of the community from fires and geologic hazards, including features necessary for such protection as evacuation routes, peak load water supply requirements, minimum road widths, clearance around structures, and geologic mapping in areas of known geologic hazards.

The Path to Passage

The origins of the Seismic Safety Element requirement lie in the very earliest meetings of the Joint Committee's Advisory Group on Land Use Planning. At one of its earliest meetings (April 21, 1970) the group reviewed its objectives in advance of preparing a June 30 progress report to the JCSS. The group reached consensus that a subcommittee should be appointed to draft a statement for the report:

> The statement is to recommend legislation making a seismic element of the general plan mandatory for all cities and counties. It is felt that this is an urgent matter that cannot await completion of the four-year legislative committee study. The subcommittee is to prepare a justification for the proposal as well as a general definition of a "seismic element," including reference to the four general types of seismic occurrence.[76]

Donald R. Nichols, a geologist on the staff of the U.S. Geological Survey, was representing the AG/LUP at a meeting of the Advisory Group chairpersons (later to become the "Coordinating Council" and finally "Executive Committee") where he reported that the land use group "has prepared a revision to the current State General Plan to take seismic safety into consideration as a mandatory element." Melville Owen, an attorney from San Francisco, also was attending as a representative of the Advisory Group on Governmental Organization and Performance (AG/GOP). He noted that this recommendation should "be studied ... and prepared for legislation in early 1971 as an indication of progress by the Joint Committee."[77]

Less than a month later (June 25, 1970), George Mader met with the AG/GOP. He noted:

> ... that the amendment need not be too specific. A generalized amendment will stimulate thought concerning specifics and [will] generate local action. A complex measure may risk failure. He points out the success of a similar general amendment: the Housing Element.[78]

The group unanimously passed a motion requesting that the JCSS prepare the needed legislation, but portending a major implementation issue, the group discussed the probable lack of local resources to perform the work demanded by such a technically oriented amendment. On September 24, Don Nichols met with the Advisory Group on Engineering Considerations and Earthquake Sciences (AG/EC&ES) during which time he briefed the members on an AG/LUP project to study a strip of land in the vicinity of Half Moon Bay (south of San Francisco) with the objective of "trying to identify general geologic and structural guidelines for application in general land use planning ... (and) these would be gross area guidelines for planners specifically, not guidelines for specific site analysis by engineers."[79]

The AG/LUP's proposal to amend the State Planning Act appeared for the first time formally on the Joint Committee's future legislative agenda on December 9, 1970 (60 days before the San Fernando earthquake). The occasion was a reception for the committee at the Faculty Club on the campus of the University of California, Berkeley. In addition to adding a "seismic element," other potential legislative measures discussed included standards for essential buildings (e.g., hospitals, fire stations and other critical buildings), prohibiting the construction of buildings "intended for human habitation on fault traces active within the last 1,000 years," and requiring "strong motion measurement devices in all major buildings." Several of these ideas are discussed in other chapters, and all began their legislative journeys within weeks after the San Fernando earthquake. There was a legislative "agenda in waiting."

Don Nichols, in comments prepared for the Joint Committee on May 13, 1971, while SB 351 was moving through the legislative process, provided further insights into both the study and local capabilities to prepare Seismic Safety Elements:

Our approach to our task is to examine current land use practices and controls in a representative area with a variety of seismic hazards. For this purpose, we have selected a strip of land from Half Moon Bay on the coast that extends eastward to San Francisco Bay. This strip includes numerous active and inactive landslides, the known active San Andreas Fault, two possibly active faults—the San Gregorio and Seal Cove—and perhaps others. It offers a wide variety of ground conditions with differing amplification characteristics ranging from solid granitic bedrock to very soft, saturated bay mud. It includes twelve incorporated communities in parts of two counties with many different building codes, zoning ordinances, and land use policies. High-rise office and apartment buildings, commercial, industrial, and single-family residential structures are present in parts of the area; other parts are unused or in agriculture and are subject to immediate and long-term developmental pressure. From our study of specific land use practices in both hazardous and non-hazardous areas, *we hope to devise legislative recommendations and criteria to guide future policies, zoning ordinances, and building regulations that can be applied statewide.*[80]

Nichols' statement also recognized the probable lack of technical data and knowledge in many communities:

We recognize that much basic data on which to base such plans does not exist in many areas. Where it does exist, however, there should be no excuse for allowing structures and human occupancy to be endangered from lack of planning. Where the information does not exist, this legislation should provide the planners with the impetus to request it from appropriate state, federal, or local agencies.[81]

On January 14, 1971, about three weeks before the San Fernando earthquake, the EC&ES group, through which flowed all legislative proposals and other recommendations, reviewed and approved the draft legislation that the Office of Legislative Counsel had prepared at Senator Alquist's request. The draft bill would add a "Seismic Safety Element" to the State Planning Act.

The AG/LUP met on January 19, 1971, and its members reviewed the draft Seismic Safety Element legislation. It was a spirited discussion, providing some insights into various aspects of the bill before it moved toward formal introduction:

> Mr. Nichols comments that the tentative draft legislation's provision is to force the education of planners when complying with it. Mr. Mader indicates a concern that the draft does not require a statement of policy as part of the seismic safety element. Mr. Cameron states that the lack of a time element may help establish a policy. Mr. Stockwell feels that the plan is an intention of policy. Mr. Blair comments that the provision can be interpreted as strongly as necessary. Mrs. Henderson states that the provision provides initial momentum for legislative proposals. Mr. Mader feels that we are heading toward policy, not appraisal. A motion is seconded and carried to accept the tentative draft as written.[82]

Two days later, the AG/GOP's members concurred in the draft language to add a sub-section (f) to Section 65302 of the Government Code related to local planning. Thus, the proposed legislation was ready to go when the proverbial window of opportunity opened: the February 9, 1971, San Fernando earthquake.

On February 10, 1971, Senator Alquist stated at a Capitol press conference that "Yesterday an aerial and ground tour of the earthquake damage area was made by members of the legislature's Joint Committee on Seismic Safety ... (and the committee's) immediate legislative recommendations include ... require Seismic Safety Elements in all city and county zoning and master planning."[83]

On February 11, two days after the earthquake, the EC&ES group met with the chairs of the other advisory groups in San Francisco. The EC&ES also concurred with the proposed wording and directed that the proposed Seismic Safety Element legislation be sent to the parent Joint Committee for immediate introduction into the legislative process, where it became Senate Bill 351 of the 1971 General Session.

Following the February 19 introduction of the bill, a closer look at the legislative process shows that SB 351 was referred on February 25 to the Senate Committee on Local Government, from which it received a recommended "do pass as amended" vote on April 21. Roy Cameron, then chairman of the AG/LUP, noted in his testimony before the committee that "The purpose of the bill is to encourage local jurisdictions to be aware of and plan for seismic safety hazards unique to their local area ... [and] there is no known opposition to the bill."[84]

With minor amendments, SB 351 on April 26 reached the Senate floor, where the bill passed the Senate and went to the Assembly. On April 28, SB 351 was referred to the Assembly Committee on Planning and Land Use for hearing on May 26, 1971. It received a unanimous "do pass" with the committee recommending placement of SB 351 on the Assembly's Consent Calendar (a procedure for quickly processing non-controversial bills). SB 351 passed the full Assembly as a consent item on June 2, and the Senate (SB 351's house of origin) consented to the bill's final wording on the following day. SB 351 landed on the governor's desk for signature at 4:00 p.m. on June 8, 1971.

The final step was securing Governor Reagan's signature on SB 351. Typically, the sponsoring author sends the governor a letter "respectfully" requesting his affirmative action. Senator Alquist's letter urged the governor to sign the bill, and the senator's letter noted that SB 351 was prepared by the JCSS' advisory "experts in the field who carefully evaluated the need for such legislation ... [and that the bill] includes no cost at the state level, nor does it impose any *appreciable* increase in costs of planning activities at the local level."[85] Governor Reagan signed SB 351 on June 17.

Early Implementation

While SB 351 "sailed" through the legislature very soon after the San Fernando earthquake, the real work to make it effective actually began before the bill's final passage. The shift in focus from the legislature to Governor Reagan's administration became the means for addressing the law's implementation.

The brevity of the Planning Act's various requirements thus meant that Governor Reagan's administration had the responsibility for promulgating detailed guidance for use by local governments on how to implement the act's several requirements. Thus, early implementation of the SSE requirement shifted quickly from the legislature to the executive branch. Senator Alquist, when he responded to a request for information from the City of Chula Vista, reflected the differences between the legislative and the executive branches: "The Joint Committee looks to the Executive Branch of the state government to implement the legislation that was passed."[86]

SB 351 fits the category of programs that are developed by the state but administered primarily by local government. Early implementation was monitored by a researcher, who noted that "local governments are where the (state seismic safety policies) must be given operational meaning.... Guidelines that indicated the kind of seismological and geological data to be incorporated into the [(Seismic Safety] Elements are published by the state. Local governments, however, actually prepare the element and then are solely responsible for any implementation." He noted further that "the success of seismic safety policy implementation by California local governments heavily influences the fate of most earthquake mitigation efforts."[87]

As SB 351 was nearing the end of its legislative journey, the members of the AG/LUP simultaneously began to hear more about potential implementation problems. Mary Henderson, a local government representative and also a member of the AG/GOP, noted that "requiring a Seismic Safety Element in general plans is not acceptable to all local governments. Restrictions are not acceptable when little information is provided on their administration." In response, Don Nichols said, "we must recognize opposition to these bills ... [and] local government has not drawn upon talented persons to assist in solving the problems."[88]

In the spring of 1972, at a Joint Committee Conference on Seismic Safety and Public Policy in Sacramento, George Mader, an AG/LUP member and a San Fernando study consultant, commented on the early implementation of the law by three cities and Los Angeles County by noting that "most general plans don't include seismic elements. The Seismic Safety Element passed in the legislature last year is not specific enough."[89] This probably was the first call for the preparation of guidelines to help local agencies prepare their Seismic Safety Elements (see below).

As noted earlier, SB 271 (Chapter 1104, Statutes of 1975) mandated Safety Elements for local general plans. The requirement for a separate SSE continued until 1978, when it was merged into the more comprehensive Safety Element. Several unsuccessful attempts were made over the years to make the SSE and some other required planning elements permissive rather than mandatory. In 1984, AB 2038 (Chapter 1009, Statutes of 1984) expanded the list of mandatory Safety Element issues, merged them with the Safety Element's requirements, and it deleted the Seismic Safety Element from the list of mandatory general plan elements. The Safety Element remains one of the requirements that all local comprehensive general plans must incorporate.[90]

When AG/LUP's members reviewed George Mader's report on the San Fernando earthquake, they showed a clear understanding of the state-local complexities involved in SB 351's implementation:

> The major problem in land use planning is determining the proper distribution of powers between the local and state levels in terms of seismic safety. Specific changes in state legislation ... can be an immediate response to improve seismic safety. The long-term response needs continuous review of state and local planning.

> The Division of Mines and Geology or the Council on Intergovernmental Relations (CIR) may be a source for such reviews (or possibly "a more centralized agency on the state level to review seismic considerations" is needed).

> Mr. Cameron [then chairman of the AG/LUP] points out that a 1971 standing committee of the legislature passed a bill out of committee requiring state review of general plans by CIR. This bill imposed such restrictive measures that local government would have been stymied in its actions. The State Office of Planning is presently charged with review of this type.[91]

When he presented his report on the San Fernando earthquake at a special hearing of the Joint Committee on February 9, 1972—the first anniversary of the San Fernando earthquake—George Mader further reflected the complex intergovernmental relations governing planning in California:

> Some rationalization of this complex situation appears necessary if the land use planning process in California is going to adequately recognize seismic safety.... A major question in all such recommendations is the appropriate division of responsibility between state and local government.... The state has not specifically urged or required local government to take on such (seismic safety) responsibilities.... On the other hand, local government has by and large not voluntarily assumed these responsibilities.

> With these observations as background, two major recommendations are made:

1. The planning process at the local level should be significantly improved with respect to seismic safety.

2. The state should assume added responsibility in providing guidance and in some cases direct control over land use decisions at the local level.[92]

Concern about the general nature of the bill's language, the time pressures of the legislative process, and little local experience and capability to address these "technical" issues led to a commitment that the AG/LUP would take the lead in defining a process that would develop implementing guidelines sooner rather than later. They were to address:

1. An effective date for adoption of the element by cities and counties.

2. A fuller description of the contents of the element.

3. Guidelines for preparation of the element.

4. Expanded geologic capabilities to collect and interpret geologic information at local, state, regional, and national levels.[93]

The result was that it would be up to the Governor's Earthquake Council (GEC), working through the Office of Planning and Research (OPR) and its former Council on Intergovernmental Relations (CIR), to provide implementing guidance to California's local governments.

Who had to prepare Seismic Safety Elements and by when?

Within a few weeks of SB 351 being signed into law as Chapter 150 of the Statutes of 1971, questions arose about which local jurisdictions were governed by the new law: Did it include both general law and charter cities? Professor of Architecture George Simonds, at a July 8, 1971, EC&ES meeting noted that if the new law applied only to non-charter (general law) cities, "it is not worth much." Sal Bianco, a JCSS contract consultant responded that after a Coordinating Council meeting on July 2 where this subject came up, the

JCSS already had requested legislative counsel to "render an opinion on this question."[94]

Legislative counsel's July 7 opinion noted in part:

> Section 65700 of the Government Code provides that the provisions of the chapter (150) shall *not apply* to a charter city, *except* to the extent that the same may be adopted by charter or ordinance of the city.... Therefore, the provisions of Chapter 150 of the Statutes of 1971 *apply only* in those chartered cities which have adopted by charter or ordinance the local planning provisions of state law.[95]

Senator Alquist quickly introduced "clean-up" legislation in 1971 and 1972 to ensure that the Seismic Safety Element requirement applied to all cities. SB 1489 of 1971 stated that SB 351 applied to both charter and general law cities, and because of continuing conflict, 1972's SB 591 reinforced 1489 by further specifying that the new law applied to charter cities.

Shortly after the July 7 opinion, consultant Sal Bianco reported to the AG/LUP that Senator Alquist had asked legislative counsel for another opinion because SB 351 "contained no deadline for local government performance"—the question being: "Are such deadlines established by reference to other sections of the Government Code?" Counsel's response was that except for two unrelated elements required by 1970 legislation:

> The local planning act ... does not contain any express deadlines for the formulation of a general plan by local agencies. Further, our research has not disclosed the existence of any general statutory deadline for adoption of a general plan containing the elements specified by Section 65302 (where the Seismic Safety Element law was placed in the codes—author).

Thus, the issue of when local governments must comply with preparing their Seismic Safety Elements had to be addressed. The issue was anticipated as early as November

30, 1971, with Karl V. Steinbrugge, now chairman of the Coordinating Council, by JCSS consultant Sal Bianco, who wrote a lengthy letter about the advisory groups' 1972 plans that said "… perhaps future legislation would establish a deadline date for counties and cities to comply with the bill's (SB 351) provisions."[96]

It was also an issue with the GEC, when, at its March 13, 1972, meeting, member Gene Block asked about the apparent delay in implementation. The minutes noted, "that legislation may be proposed by the League (of California Cities) to delay implementation of this and other recently mandated general plan elements."[97]

Speaking to the California Civil Defense and Disaster Association in Bakersfield on May 21, 1974, JCSS consultant Brynn Kernaghan reported that Senator Alquist had introduced legislation (SB 2365) to require OPR and CIR to review general plan elements. The Seismic Safety and Safety Elements would be forwarded to the Division of Mines and Geology (DMG) for "special checking."

Local capabilities—the struggle for guidance

Without question, the biggest challenge following enactment of SB 351 was rapidly providing useful guidance to California's local governments to help them prepare their first Seismic Safety Elements. The preparation of implementing guidance went through several stages over a period of about three years before it became fully institutionalized in the Office of Planning and Research (OPR).

Reflective of communities' concerns are two early letters to the JCSS, one from the Southern California City of Arcadia and the second from the Council of Fresno County Governments. The acting planning director of Arcadia said, "I am interested in finding out what material and/or assistance this committee (the JCSS) could provide to local jurisdictions, in particular, the City of Arcadia."[98] Speaking for all of Fresno County's local governments, the Council's letter referred to its consideration of an area-wide approach (the county and 15 cities), but:

> Our problem is that we have no idea as to the extent and probable cost of a Seismic Safety Element; or the availability of technical and financial assistance from the state and federal governments, if any. We would appreciate any help you can give us in this respect.[99]

Yet, some localities did move ahead, probably because they were aware of their seismic risk and had taken some steps to mitigate it. The City of Hayward, through which the notorious Hayward Fault runs, thanked the JCSS for its support:

> On October 25, 1972, the City Council adopted the Hayward Earthquake Study as the Seismic Safety Element of the City's Master Plan. Proposals contained therein are thus City of Hayward policy, and the proposed procedures for implementation should be carried out as part of the ongoing work program of City departments and other concerned individuals.

> On behalf of the Planning Department, I would like to express the appreciation of our staff for your participation and continued assistance on this program.[100]

From a broad state perspective, however, things had to move quickly, and members of AG/LUP found themselves working hard with state agency representatives, initially through the Governor's Earthquake Council (GEC) and later through other mechanisms, to provide the needed implementation "guidance" for local planners.

About 90 days after SB 351's enactment, George G. Mader of the AG/LUP provided the context and some ideas for the content of an "ideal" Seismic Safety Element, when he spoke to the 73rd Annual League of California Cities Conference:

There has been little experience in this country in developing such an item as a seismic safety element. A considerable amount of experience will be necessary within and between the professions having an interest in seismic problems—soils engineers, engineering geologists, seismologists, structural engineers, lawyers, planners, etc.—before an adequate description of the general content and form of a seismic safety element can be developed. Nonetheless, there has been sufficient experience to allow some tentative suggestions for the content of a Seismic Safety Element. Of course, the specific content will vary from area to area depending on particular seismic problems.

First, I would suggest the element contain a diagram and accompanying text which set forth the relative risks involved in developing the different lands within a jurisdiction. The diagram should probably delineate in broad risk zones the limitations proposed for land uses. For instance, the diagram might define a fault zone where the risk is very high and state that only open space uses will be permitted. A band might be shown next to the fault zone where the risk is less and uses should not be more intense than, for example, one-story single-family frame residences. Another example of a risk area would be an existing landslide. The diagram might indicate that on existing landslides no development or only limited development should be permitted unless the landslide can be stabilized. In areas of potential instability, the diagram might propose developments that would require minimum grading and permit only light structures. The result would be a diagram which is not a plan but rather a delineation of constraints on development. This diagram would provide one basis for the preparation of the other elements of the general plan, such as the land use and open space elements.

The diagram might also indicate those areas requiring further geologic investigation before planning or development decisions can be made. In this way the element would take on a programmatic aspect and guide the jurisdiction in its geologic mapping and planning and development programs.

A second part of the element should consist of recommendations for implementation. While broad geologic risk zones can probably be defined at the general plan level and may be adequate for making broad decisions, much more detailed information will be needed at subsequent stages of the planning and development process. The implementation program should include recommendations for modifications to the zoning ordinance which will recognize seismic constraints. These might include recommendations for such items as specific zones along fault traces or special zoning of areas subject to landsliding or other ground failures. The program should also include recommendations for the submission of adequate geologic information as a part of applications for subdivisions, grading permits, and building permits. Geologic input is extremely important as part of these procedures. It is during the subdivision and grading periods of development that detailed geologic data will be needed. As an aside, it should be pointed out that for any of these regulations to work effectively at a local level, the jurisdiction must have appropriate geologic capabilities either on its staff or available on a consultant basis to advise the community of the adequacy of the data submitted. The requirement is very similar to the practice of jurisdictions having their own engineers to review the work of engineers retained by developers.

The implementation program should also include recommendations for the review of public improvement projects with respect to geologic information. And finally, the program should recommend that the building code include requirements adequate for the types of seismic hazards anticipated.[101]

How, by whom, and by when would the implementation guidance be prepared? In early September 1971, when noting that SB 351 had been approved by Governor Reagan, JCSS consultant Sal Bianco reported to the JCSS Coordinating Council:

The implementation of such a measure was considered by [the] AG/LUP when it [SB 351] was drafted, but his [Mader's] group felt the bill's impact would be in the area of making local communities aware of the need to consider seismic safety in their planning. Perhaps the Division of Mines and Geology should play a role in developing background material to assist local jurisdictions in preparing a seismic element for their general plans?[102]

Initially, "off the shelf" materials were distributed by the JCSS' consulting staff to local government agencies that requested assistance, and in early 1972 the Council on Inter-Governmental Relations "expressed interest in preparing criteria for a Seismic Safety Element."[103]

An Ad Hoc Group on the State Providing Technical Expertise to Local Governments was formed as a multiagency and intergovernmental group under the auspices of the JCSS. Its purpose was to address issues associated with the need to ensure local technical capability to effectively implement both SB 351 and SB 520 (Geologic Hazards Zones Act of 1972). The group monitored related legislation in process at the time, including several bills related to planning and land use. The minutes for one of the group's 1974 meetings note that:

> A bill will be introduced or amendments will be offered to an existing bill to:
>
> 1. authorize the Office of Planning and Research to review the elements of the General Land Use Plans of local jurisdictions;
>
> 2. authorize the Department of Conservation to review the seismic safety and safety elements of the general plan; and
>
> 3. appropriate the sum of $50,000 to the Department of Conservation for purposes of this act.[104]

Guidance Emerges

While the JCSS consulting staff was sending out useful materials when requested by local agencies, AG/LUP members Mader and Nichols, especially, began working with the CIR and through it the Governor's Earthquake Council (GEC) to prepare more complete guidelines for local use. In October 1972 the AG/LUP approved a set of "working guidelines," which Senator Alquist transmitted for review to the CIR. Simultaneously, the GEC had drafted a set of guidelines that the CIR distributed to local governments, largely because CDMG had received "numerous requests for assistance," which it met by also providing readily available materials.

CDMG's draft "interim guidelines" emerged in May 1972. CDMG established the following recommended priorities:

A. Policy statement regarding seismic hazards

B. Emergency plan and organization for earthquake response

C. Abatement of existing structural hazards

D. Identification, delineation and evaluation of local seismic risk

E. Reducing seismic risk to future developments and structures

F. Provision for updating components of the seismic safety element to incorporate new data

James G. Stearns, GEC chairman for Governor Reagan, distributed copies for review and comment to two of the council's committees: Preparedness and Response, and Research and Investigations. Among Stearns' major points were:

No provision was made in the bill (SB 351) for the development of state criteria and guidelines as to what should comprise that element.... Inquiries from many cities and counties to several state agencies have been received.... Several studies are in progress or about to commence and ... these should result in the development of a model element.... The Governor's Earthquake Council is ideally constituted to prepare interim guidelines for the Seismic Safety Element ... [and] these can be incorporated into a final document for distribution to local governments by the State Council on Intergovernmental Relations before July 1.[105]

Following extensive comments among the GEC's members, in July 1972 the CIR published and distributed its "Seismic Safety Element Interim Guidelines" to all of California's local governments. The document explained the origin and scope of the new law, provided technical definitions and descriptions of damage-causing factors and types of losses, and with slight wording changes and supporting appendices retained the above-suggested contents of a Seismic Safety Element.

Local reaction to the guidelines was strong and not always complimentary. Excerpts from "notes" provided by a staff member of Contra Costa County's Planning Department to the ad hoc committee in July 1972 and later (September) to George Mader and Don Nichols provide insights:

> The "guidelines" overlook the purpose and character of a general plan—which is a jurisdiction's most basic and general long-range policy instrument—and address themselves mainly to succeeding instruments in the plan hierarchy, especially disaster response and recovery organization and operations.[106]

The notes then offered a detailed outline of a "model element" that fit into the typical local general planning processes. It provided a "recommendations" section for the element that would address the seismic safety of new developments and structures, existing structures, essential facilities, major buildings and places of assembly, uses of hazardous areas, and post-earthquake economic recovery and redevelopment.

Work started virtually immediately on revising the guidelines. Some of the GEC's members continued to be concerned about what was included in the guidelines, and Senator Alquist, on behalf of the AG/LUP, wrote to the CIR, noting:

> We ... have considerable interest in pursuing implementation of this legislation.... The suggestions of the Advisory Group should be in a form suitable for discussion within the next several weeks.... Following that, members (of the AG/LUP) would like to arrange a meeting with you and other interested parties to discuss the guidelines.[107]

78

Several documents were prepared during the 1972–4 period to support SB 351's early implementation. William J. Petak, then an Associate Professor of political science at California State University, Fullerton, prepared a "Procedure for Developing a Seismic Safety Element for the General Plan" (June 1972). This document was to provide background information before the GEC published its interim guidelines. It was presented at a June 1972 meeting of the Planning Directors Association of Orange County, and Professor Petak submitted it to the American Society of Planning Officials (ASPO) and the American Institute of Planners (AIP) for possible publication.

The AG/LUP also prepared draft "Guidelines of the Preparation of a Seismic Safety Element" in September 1972. The covering note referred to both the GEC's Interim Guidelines and Professor Petak's paper, but also noted that the Council on Intergovernmental Relations had been "charged by the legislature with devising and adopting *official guidelines* for all elements of the General Plan. Interestingly, these guidelines recommended a sequence of activities and development of a policy statement, and suggested ways to address the law's specific seismic hazard requirements: fault displacement, ground shaking, ground failure, and tsunami and seiche effects.

Soon thereafter, OPR became the source of official general planning guidelines. OPR's 2003 guidance notes:

> The aim of the element is to reduce the potential risk of death, injuries, property damage, and economic and social dislocation resulting from fires, floods, earthquakes, landslides, and other hazards. Other locally relevant safety issues, such as airport land use, emergency response, hazardous materials spills, and crime reduction, also may be included. Some local jurisdictions have even chosen to incorporate their hazardous waste management plans into their Safety Elements.[108]

Major Policy Issues since Enactment

My review of legislative records between the charter law's date of enactment (June 17, 1971) and 2001 shows about 15 successful or defeated attempts have been made to amend this law. Some of the most prominent issues that arose during this period (and sometimes multiple times) included establishing the deadline for completing the element; strengthening the state's role in reviewing and approving Seismic Safety and Safety Elements; making the requirement permissive rather than mandatory; allowing cities and counties to prepare combined Safety Elements; requiring the Seismic Safety Commission to study and report to the legislature on building codes and standards governing the design and construction of new buildings and critical facilities; addressing issues of existing earthquake-hazardous buildings; making the entire State Planning Law discretionary; and granting exemptions from all planning requirements for up to five new communities.

George Mader noted in 2009 that "OPR's latest general plan guidelines contain eight pages related to Safety Elements. The point is that over time, experience in preparing the seismic portions of Safety Elements has provided guidance that was omitted from the element when first introduced. As noted in early presentations on the element, experience would in effect produce guidelines. This is essentially what has happened."[109]

Tinkering with the organization of California's state government—the establishment of the Seismic Safety Commission—is the focus of the next chapter: Wave Two, Chapter 5.

ENDNOTES

73. California Legislature, Special Subcommittee of the Joint Committee on Seismic Safety, July 1972, *The San Fernando Earthquake of February 9, 1971 and Public Policy*: 73.

74. Alfred E. Alquist, press release, February 19, 1971.

75. George G. Mader, "Thoughts on the New Seismic Safety Element Requirement of the California Planning Law" (paper, 73rd Annual League of California Cities Conference, September 27, 1971).

76. Advisory Group on Land Use Planning, minutes of April 21, 1970: 1.

77. Advisory Group on Engineering Considerations and Earthquake Sciences and Advisory Group Coordinating Council, joint meeting minutes of May 28,1970: 2.

78. Advisory Group on Governmental Organization and Performance, minutes of June 25, 1970: 2.

79. Advisory Group on Engineering Considerations and Earthquake Sciences minutes of September 24, 1970: 2–3.

80. Donald R. Nichols, statement prepared and presented on May 13, 1971, to the Joint Committee on Seismic Safety, unnumbered pages, emphasis added.

81. ibid.

82. Advisory Group on Land Use Planning, minutes of January 19, 1971: 3.

83. Alfred E. Alquist, press conference statement, February 10, 1971.

84. The last phrase is an important clue to legislators about the level and sources of conflict involved, important considerations to their voting—author.

85. Alfred E. Alquist, letter to Governor Ronald Reagan, June 7, 1971, emphasis added.

86. Alfred E. Alquist, letter to Lawrence M. Yamagata, City of Chula Vista, May 1, 1973: 1.

87. Alan J. Wyner, "Implementing Seismic Safety Policy: The Case of Local Governments in California," in *Social and Economic Aspects of Earthquakes*, eds. Barclay G. Jones and Miha Tomazevic. Proceedings of the Third International Conference: The Social and Economic Aspects of Earthquakes and Planning to Mitigate Their Impacts (Ithaca, New York: Cornell University, 1982): 288.

88. Advisory Group on Land Use Planning, minutes of May 18, 1971: 2.

89. Joint Committee on Seismic Safety, summary notes (Conference on Seismic Safety and Public Policy, May 10–12, 1972): 10.

90. Governor's Office of Planning and Research, *General Plan Guidelines*, 2003: 90.

91. Advisory Group on Land Use Planning, minutes of January 18, 1972: 2.

92. George G. Mader, statement for the February 9, 1972, hearing of the Special Subcommittee of the Joint Committee on Seismic Safety to Investigate the San Fernando Earthquake of February 9, 1971: 1–2.

93. California Legislature, Special Subcommittee of the Joint Committee on Seismic Safety, *The San Fernando Earthquake of February 9, 1971, and Public Policy*, (Sacramento, July 1972): 81.

94. Advisory Group on Engineering Considerations and Earthquake Sciences, minutes of January 8, 1971: 5.

95. Legislative Counsel, Opinion #15095 to Senator Alfred E. Alquist, July 7, 1971, emphases added.

96. Letter to Karl V. Steinbrugge from Sal Bianco, November 30, 1971: 2.

97. Governor's Earthquake Council, minutes of March 13, 1972: 4.

98. Letter to Alfred E. Alquist from the City of Arcadia, October 15, 1971.

99. Letter to Rodney J. Diridon from the Fresno County Council of Governments, December 22, 1971.

100. Letter to Brynn Kernaghan from Bruce P. Allred, November 6, 1972.

101. George G. Mader, "Thoughts on the New Seismic Safety Element Requirement of the California Planning Law," William Spangle Associates, Portola Valley, September 27, 1971, 2–3.

102. Advisory Group Coordinating Council, minutes of September 2, 1971: 4. The Division of Mines and Geology later requested one year to develop a "model seismic safety element"—author.

103. Advisory Group on Disaster Preparedness, minutes of March 15, 1972: 2.

104. Ad Hoc Group on the State Providing Technical Assistance to Local Governments, minutes of March 26, 1974.

105. James Stearns, memorandum to Herbert Temple and Wesley Bruer, May 22, 1972.

106. Notes on "Suggested Interim Guidelines for the Seismic Safety Element in General Plans," May 21, 1972. Author unknown.

107. Alfred E. Alquist, letter to James R. Johnson, October 6, 1972.

108. Governor's Office of Planning and Research, *General Plan Guidelines*, 2003: 90.

109. George G. Mader, memorandum to Bob Olson, May 1, 2009.

WAVE TWO, CHAPTER 5:
Advocates To The Political Table—
The Seismic Safety Commission Act
Of 1974

Unless the state ... acts quickly and with decisiveness to formulate a public policy position minimizing the loss of life and property, we face a monumental disaster.[110]

This chapter traces the multi-year political process that led to a "bureaucratic innovation" by addressing a critical seismic safety policy question: Why and how did the state's political process work to create a "state earthquake agency?" It took knowledge, advocates, events, and political leadership.

Preludes to Legislative Action

The Joint Committee on Seismic Safety (JCSS) was given a four-year life (1970–74) to issue a report to the legislature. It was very interesting to see how early discussions then began among the JCSS' advisors about a follow-up on a more "permanent" effort of some kind.

The resulting 1974 Seismic Safety Commission Act emerged from a "messy" four-year political process. The law was the carefully negotiated product of ideas that can be traced to learning from "relevant" damaging earthquakes; a 1967 report to the Secretary of the Resources Agency; a 1968 monograph titled *Earthquake Hazard in the San Francisco Bay Area: A Continuing Problem in Public Policy*; the 1969–74 work of the approximately 70 volunteer advisors to the legislature's Joint Committee on Seismic Safety (JCSS); the 1972–74 work of Governor Reagan's Earthquake Council (GEC, formed following the 1971 San Fernando earthquake); one legislative proposal for a "Public Safety Agency," another for a "Hazardous Buildings Commission," two others to form executive branch "earthquake offices"; and, finally, San Francisco Bay Area models of regional planning and governance, such as the Bay Conservation and Development Commission and the Metropolitan Transportation Commission.

Interestingly, even after the 1933 Long Beach earthquake a proposal emerged to create a regional "earthquake body" for the Los Angeles area. It would have been composed of "the presidents of all the Chambers of Commerce in Southern California, presidents of all the technical organizations and leading civic associations with the avowed object of promoting public safety against seismic hazards."[111]

However, it took the February 9, 1971, San Fernando earthquake to create a more receptive political climate for proposing a commission, which might not have been possible otherwise. Many of the factors, especially the JCSS' advisors' internal "debates" about long-range seismic safety policy and a mechanism for addressing it, existed from the committee's beginning in 1969. The discussions just became more serious and focused after the earthquake when optimism about creating something appeared more politically feasible. For example, two days afterward, the JCSS' Advisory Group on Engineering Considerations and Earthquake Sciences (EC&ES) reviewed a list of potential legislation, including the "establishment of regional boards patterned after the BCDC (Bay Conservation and Development Commission)

Engineering Criteria Review Board to review major building proposals of government and public agencies."[112]

Learning from Earthquakes

Academic and practicing earth scientists and structural engineers furthered their interest in earthquake engineering by forming the Earthquake Engineering Research Institute (EERI) in 1949. Its members often visited and reported on the effects of damaging earthquakes. Several "relevant" ones included the Arvin-Tehachapi, California (1952), Alaska (1964), Niigata, Japan (1964), and the Caracas, Venezuela (1967) earthquakes. The February 9, 1971, San Fernando, California, earthquake provided especially important knowledge that ultimately led to the Seismic Safety Commission's formation.

Professional meetings, conferences, and publications followed these earthquakes, all of which helped increase knowledge about the performance of buildings, utility and transportation systems, and other structures when subjected to earthquake motions. This learning experience also served to raise concern about California's vulnerability to similar events and helped create a consensus about what "should" be done (from scientific and technical viewpoints) to reduce the state's risk.

For example, Joint Rules Committee Resolution No. 7 of April 27, 1971, gave $150,000 from the legislature's contingency funds to the JCSS to "write a detailed report on the San Fernando earthquake, which would include suggestions for legislation."[113] This report plus those of the individual advisory groups led to the appointment of 23 ad hoc groups in 1974, each of which was charged with drafting individual pieces of legislation to further the implementation of the JCSS' four years of work even while the advisors were finishing the joint committee's final report.

Earthquake and Geologic Hazards in California: A Report to the Resources Agency

This largely internal 1967 document titled *Earthquake and Geologic Hazards in California: A Report to the Resources Agency* was a consequence of two post-Alaska earthquake conferences sponsored by the Resources Agency (largely the Division of Mines and Geology and the Department of Water Resources and its Division of Safety of Dams) in 1964 and 1965. The report tied California's vulnerability directly to the Alaskan experience:

> The great Alaska earthquake of [March 27] 1964 with resulting tsunami damage in Crescent City [California], and also the numerous landslides that have caused damage and destruction of buildings and homes in both the northern and southern parts of the state during recent years (especially the 1964 floods) are pointed reminders that California faces hazards deriving from geologic conditions.[114]

The conferences, one in San Francisco and another in Los Angeles, were attended largely by practicing, government, and research engineers and earth scientists. Only a few of those attending the San Francisco session had public policy interests:

> Stan Scott [of UC Berkeley's Institute of Governmental Studies], Richard (Bud) Carpenter from the League of California Cities, and Bill MacDougall of the County Supervisors Association [attended]. Asked to speak ... during the closing remarks, Bill McDougall agreed something had to be done to mitigate earthquake hazard but added, "Whatever you do, leave it to local government."[115]

Reflecting on his lunch conversation with Bud Carpenter during the conference, Scott recalled Carpenter saying:

> This problem is really serious, but nothing is going to happen until they [scientists and engineers] get the policy people into it. They don't have policy people here. You're from IGS, and there are MacDougall and me, but we're the only policy

people here. All the rest are technical. Until they get the policy people in, nothing will happen.[116]

Governor Reagan's Secretary of Resources, Hugo Fisher, followed up the conferences by appointing two committees: the Geologic Hazards Committee for Program and another for Organization. The job of the first was to recommend programs and the second was to address implementation. Many of the committees' members included future advisors to the Joint Committee on Seismic Safety, members of the Governor's Earthquake Council, and later, the Seismic Safety Commission. The committees' report noted that "both committees emphasize the importance of facing up to the problems of earthquake and other geologic hazards, as well as the desirability of the state taking the initiative in coping with the problems."[117]

One recommendation called for creating an "Earthquake and Geologic Hazards Board" to oversee several functions:

> The planning, organizing, and implementing of the [new and expanded] programs recommended in [this] report will be a major long-term effort requiring knowledgeable and perceptive guidance on a continuing basis.
>
> To provide the required guidance and judgment, we therefore recommend establishment of a board of reputable and knowledgeable persons, not employees of state agencies, who would be consulted on all state earthquake and other geologic hazards programs.... We recommend that this State Earthquake and Geologic Hazards Board be composed of twelve to fifteen individuals of recognized competence and reputation in civil engineering, geology, seismology, and other pertinent technical and scientific fields, and that they be appointed by the governor and serve without remuneration. The board should have the services of a full-time paid executive secretary and office staff.... It would be the function of the board to keep itself informed concerning earthquake and geologic hazards and what the state is doing about them, to advise, approve, and coordinate research programs for state agencies, to recommend programs to the legislature

for implementation and financing, to assist in obtaining funds for research on geologic hazards, and to make contracts for research with state agencies, universities, and private organizations.[118]

The report reflected some tension about this new board's proposed responsibilities and where it would fit in the bureaucracy, with the committees stating that "we do not recommend the formation of a single agency to undertake all of the work pertinent to earthquake and geologic hazards."[119] The GEC successfully negotiated a similar safeguard into SB 1729 of 1974 regarding the Seismic Safety Commission's responsibilities (see below).

The Earthquake and Geologic Hazards Board was not created, but noted Caltech engineering professor George W. Housner, who chaired the organization committee, observed later that "this recommendation may have provided a seed that later became the Seismic Safety Commission."[120] However, an existing board was broadened and renamed the State Mining and Geology Board. Several of the Geologic Hazards Committee's members were appointed to it, effectively forming a subcommittee to address their earthquake and geologic hazards interests.

Earthquake Hazard in the San Francisco Bay Area: A Continuing Problem in Public Policy

Stanley Scott, Research Political Scientist at UC Berkeley's Institute of Governmental Studies (IGS), who also became interested in the policy aspects of seismic safety because of the Alaskan earthquake, prepared a bibliography on earthquake policy issues in 1965 following his attendance at the 1964 San Francisco conference. He decided to pursue the subject further because he sensed a "policy void," and "it was that conference that got me interested in our institute doing something on the public policy aspects of earthquake hazard and earthquake preparedness."[121]

It was during a conversation with Ian Campbell, then the State Geologist, that Scott was referred to Dr. Perry Byerly, Emeritus Professor of Seismology on the UC Berkeley campus. Byerly discussed the idea of a monograph with Dr. Bruce Bolt, who had joined the faculty about two years earlier. Bolt suggested Karl V. Steinbrugge, and through Byerly, Stan Scott contacted Karl,[122] who was a structural engineer and part-time faculty member of UC Berkeley's Department of Architecture. Karl spent considerable time during his career investigating and reporting on significant damaging earthquakes, including the 1964 Alaskan earthquake.

Stan Scott then approached Karl about writing a monograph about earthquake hazards in the San Francisco Bay Area. It was to become one of the Institute's "reports on public policy problems and issues confronting the San Francisco Bay Area" because residents and local and state officials alike "have only begun to think about the governmental implications and public policy needs posed by the region's serious and omnipresent—but silent and unseen— earthquake hazard."[123] Scott later recalled that Steinbrugge "was commissioned with the express purpose of alerting policymakers and the public to the continuing earthquake hazard in the San Francisco Bay Area, and suggesting what might be done to improve seismic safety."[124]

The result in 1968 was the monograph titled *Earthquake Hazard in the San Francisco Bay Area: A Continuing Problem in Public Policy*. Karl's (and collaborator Stan Scott's) two policy-oriented recommendations were modest: "Planning for best land use where earthquake geologic hazard may exist" (in and across active fault zones, on bay fill lands, and near potential landslide areas) and "reducing the life hazard of older non-earthquake-resistive, collapse-hazard structures."[125]

More importantly from a policy perspective, however, was the author's discussion of how seismic safety might be addressed by "some type of regional government."[126] This idea fit in well with the monograph series' focus and local and state efforts to address regional issues:

> One organizational approach to regional land-use problems would assign them to some type of regional government. To be effective, such a regional government would have to be adequately staffed and funded, and given sufficient authority to act in cases where earthquake geologic hazards exist. A regional governmental organization would probably be better able to resist political pressures that might otherwise influence the leaders of a small community, especially one encompassing large undeveloped land areas in earthquake geologic hazard zones.
>
> A regional government would also have enough earthquake geologic hazards work to justify an adequate staff. Technical standards involving several professional and scientific disciplines would have to be established. These standards must be the work of a professional and scientific group, and not that of an individual, since broadly based and sound judgment is very important. The professional and scientific group should serve as consultants, or as an advisory board. The group should include geologists, soils engineers, city planners, structural engineers, business economists, and others. Their jurisdiction should include publicly owned utilities, as well as private property.
>
> In conclusion, the principal earthquake *geologic* hazards in the Bay Area are significant regional land-use problems best solved by some type of regional government. This government must have adequate funds and authority to plan, and to implement its plans, using a competent professional interdisciplinary approach. The *nongeologic* earthquake hazards, principally older non-earthquake-resistive buildings, have been and probably should continue to be the responsibilities of the cities and counties.[127]

In a follow-up article, Scott recommended the creation of an "earthquake policy commission" for the San Francisco

Bay Area.[128] Policy advocacy was emerging, and these ideas led to subsequent legislative proposals to create a regional or statewide seismic safety commission, the composition of the five advisory groups to the Joint Committee on Seismic Safety, and later, the Seismic Safety Commission.

The state's political context also was changing at the time. According to one observer:

> 1968 was a big year of change in the California state legislature. The November elections saw twenty former state assemblymen elected to the State Senate; 1968 was also a year of reapportionment of the state's congressional boundaries and the legislators' districts. In addition, 1968 was only the second year of full-time legislative sessions. Prior to 1966 California had a part-time legislature. More items were being addressed at the state level. Prior to his election in 1966, Alfred Alquist served in the State Assembly. Senator Alquist was re-elected to a four-year term in 1968 and hired Steve Larson as his first full-time staff aide.
>
> Steve was a native of San Jose and had members of his family who remembered the devastation of the 1906 San Francisco earthquake. He [Larson] says that he, like almost everyone else, assumed earthquakes were something we were "locked into suffering through."[129]

The IGS routinely distributed its publications to all San Francisco Bay Area legislators, including Senator Alquist, where it caught Steve Larson's attention. On March 17, 1969, about 20 invitees (including me) assembled for dinner at UC Berkeley's Faculty Club to discuss how to follow-up on Karl Steinbrugge's monograph. Some time prior to that dinner Steve called Karl Steinbrugge, and according to Stan Scott's recollection about a conversation he had with Karl:

> Karl hit the panic button, being of two minds about your [Larson] attending, saying, "Hey, what about this guy? He's from the legislature, you know. Should he be invited? What do you think? So we really conferred on that. Then we decided, "If we have a possible avenue to Sacramento, we'd better encourage that." So you [Larson] were invited.[130]

Steve returned to Sacramento, where he convinced Senator Alquist to support the seismic safety issue, with Steve recalling the Senator saying, "Well, it can't hurt me to go down this road for a while. Let's try it and see what happens."[131] In conjunction with his other staff duties, such as visiting the sites of college campus civil rights and anti-Vietnam war riots at the time, Steve's usual pattern was to visit Karl Steinbrugge in Berkeley or at his offices in San Francisco and "listen to what he had to say, and then I'd visit a riot."[132] Three legislative initiatives were taken within days after the dinner: Senate Bill 1207, Senate Bill 1041, and Senate Concurrent Resolution 128, each of which is discussed below.

L. Thomas Tobin, the Seismic Safety Commission's third executive director, observed in 1993 that:

> Those present (at the 1969 dinner) were from engineering and architecture, geology and land-use planning, emergency preparedness and local government, public policy and politics. It was a magic brew that led to a serendipitous night and the institutionalizing of earthquake safety in California and the United States.[133]

About the same time (May 1969) the Regional Office of the Federal Office of Emergency Preparedness— OEP, now the Federal Emergency Management Agency (FEMA)—sponsored a conference, "Geologic Hazards and Public Problems" (for which I was the project manager). The program linked scientific and engineering hazards information to possible public policy needs. This event, partly triggered by the 1969 Santa Barbara Channel oil spill, followed up an earlier (June 1967) internal federal seminar that explored what demands might be placed on the federal government when a repeat of the 1906 San Francisco earthquake occurred.

It was during the conference's second day that IGS' Stanley Scott first publicly mentioned Senate Bill 1207's concept of a "Seismic Safety Commission" in his talk titled "A Model Seismic Safety Program for the Bay Region." Stan reported that SB 1207 had been killed earlier that very day by a Senate committee (see below).

Karl Steinbrugge also put forward the concept of a Bay Area (or possible statewide) commission at the same meeting:

> It is proposed that a commission consisting of twenty-five persons be created by the state legislature. A total of twenty of these persons would be selected for their experience and ability to form and implement broad policy based on the recommendations from five subcommissions, which are also to be established by the legislature. The chairmen of the five subcommissions would also be members of the commission.
>
> A total of five subcommissions would be created, which would establish and control the detailed requirements in specified subject areas. Possibly well over 100 persons would be involved.... The importance of the five subcommissions cannot be overestimated; their areas of interest and their membership composition are outlined ... engineering considerations and earthquake sciences ... disaster preparedness ... post-earthquake recovery and development ... land use planning, [and] government organization and performance.[134]

It was no accident, therefore, that Senate Concurrent Resolution 128, which had been introduced prior to this conference, creating the Joint Committee on Seismic Safety designated five advisory groups with the same titles. Formal creation of a commission had to wait until 1974.

The Legislature Forms the Joint Committee on Seismic Safety

Within days of the Faculty Club dinner, Senator Alquist announced on April 3, 1969, that he would be introducing legislation to create "a twenty-seven-member Bay Area Seismic Safety Commission composed of representatives of the public, and federal, state, and local governing agencies."[135] SB 1207 was coauthored by his San Jose colleague, Assemblymember John Vasconcellos. It did not progress far, being held [killed] after its first hearing by the Senate's Governmental Efficiency Committee. SB 1207 died with 18 other bills at the end of the session on August 8 because "no committee action was taken."[136]

Virtually simultaneously, San Francisco Senator George Moscone, with other Bay Area and Los Angeles coauthors, introduced SB 1041. This legislation, worded almost the same as SB 1207, would have created a statewide Seismic Safety Commission. It met virtually the same fate in the same place when the Governmental Efficiency Committee recommended SB 1041 be assigned to "interim study."[137]

We understood that SB 1207 and SB 1041 were "long shots," and when they were assigned to the Governmental Efficiency (not the Governmental Organization) Committee, their futures were dim. Dubbed a "legislative graveyard," and chaired by Senator Richard Dolwig (not an Alquist ally), and the fact that the bills were not supported by Governor Reagan's administration, Senate Concurrent Resolution (SCR) 128 became the fallback, and likely most achievable, proposal.

SCR 128, the third measure proposed to address California's earthquake risk, would be for the legislature itself to establish a limited-term study committee. Needing agreement only from members of the Joint Rules Committee, SCR 128 created the eight-member (four from each house) Joint Committee on Seismic Safety (JCSS) on July 25, 1969. Joint Committees are temporary, funded by the legislature, and have little authority, but they are mechanisms to study and report on policy issues. Although joint committee members do not have jurisdiction to review and vote on legislation, they can and often do introduce bills based on the committees' work.

SCR 128, modeled as noted on the failed legislation to create a Bay Area Seismic Safety Commission, was criticized for its apparent parochial focus. The resolution provided that the JCSS could study "so much other of the state as it should appear advisable." We were reluctant to tinker with the only successful legislation we had; thus, I responded for Senator Alquist to a letter he had received:

> The implementation of SCR 128 and the consequent organization of the committee itself and of the advisory groups reflect a statewide concern. For example, of the eight legislators, four are from Southern California, and the membership of the advisory groups provides for strong participation from your area (Los Angeles).... We feel we are statewide in our orientation and that the products of our work will be applicable on that basis as well.[138]

Many of us involved in these legislative efforts considered SCR 128's passage a major victory. About 70 volunteers were recruited to serve on the five advisory groups established by SCR 128, and after we became organized, our objective was to prepare a report for the legislature by June 30, 1974. In 1970 the small informal "Coordinating Council" was formed, which was followed on December 8, 1971, by a formal "Executive Committee" consisting of Karl Steinbrugge and the chairpersons of each advisory group (of which I was one). It was to manage the JCSS' increased workload, especially the preparation and review of proposed legislation.[139]

Largely because of the February 9, 1971, San Fernando earthquake, Joint Rules Committee Resolution 7 of April 27, 1971, appropriated $150,000 of the legislature's own funds to further the JCSS' work, including preparing the study noted below. Senate Concurrent Resolution (SCR) 77 extended the JCSS' life through 1974 (the end of the two-year legislative session). Assembly Concurrent Resolution (ACR) 44, also passed soon after the earthquake, directed the JCSS to conduct a special study of the earthquake's policy implications, which it completed in 1972 as *The San Fernando Earthquake of February 9, 1971, and Public Policy*.

The shift in priorities—from study to advocacy—occurred quickly after the February 9, 1971, San Fernando earthquake. On February 11, two days after the event, the EC&ES members met in Los Angeles with Senator Alquist because the group would be discussing several legislative ideas of "immediate concern," and "it is our responsibility to decipher this warning, [and] your recommendations will then be the basis for legislation which will minimize the effects of our next damaging earthquake."[140] Less than a month later (on March 2, 1971), Karl Steinbrugge told the Advisory Group chairpersons that proposing legislation should be of immediate concern to each group. Special procedures were established to ensure thorough internal review of all legislative proposals before they were formally submitted to the JCSS' legislative members for their possible sponsorship.[141]

Reaching Consensus within the JCSS was Difficult

Even with the opportunities created by SCR 128 and the demands on the JCSS' volunteer advisors, the question of what, if anything, should follow the committee's final report to the legislature was discussed virtually from the beginning. Some members, such as Stanley Scott and Karl Steinbrugge, had given a long-term organizational objective some thought, and two early legislative proposals had failed already.

Prophetically, soon after the JCSS was appointed, but only about 90 days before the San Fernando earthquake, Karl Steinbrugge asked each advisory group to think about potential legislation "to have on hand either for

97

introduction during the next session or in the event that a damaging earthquake does take place. It is felt that, should an earthquake occur, the committee's reputation could suffer if it offered no tangible evidence of its work."[142]

On February 11, 1971, just two days after the San Fernando earthquake, EC&ES reviewed its list of potential legislation, including "establishment of regional boards patterned after the BCDC Engineering Criteria Review Board to review major building proposals of government and public agencies."[143] Some of EC&ES' members had served on BCDC's board and believed it had been effective.

EC&ES, being "more equal" than the other four advisory groups, received a combined package of legislative proposals on March 11, 1971. The list included six "ready to go" proposals and seventeen others that needed further development. The ready six included mandating that a "Seismic Safety Element" be added to the State Planning Act; that new hospital buildings meet Field Act standards or similar qualifications but not be retroactive; that new public emergency services buildings, such as fire stations, communications centers, and police stations, also be required to meet Field Act standards but also not be retroactive; that the Field Act itself be amended to require a geologic investigation of prospective sites for additions to existing or new school buildings; that construction be prohibited along the "actual trace of any active fault"; and that there be a statewide requirement for including strong-motion instruments in various types of buildings and locations.[144]

The other seventeen items included requiring that building inspectors be licensed civil engineers, emergency response plans be required of all local governments, regional seismic design review boards be created, seismic safety be considered in renewal and land reuse plans, changes be made in the state's subdivision regulations regarding development in hazardous areas, emergency action functions and checklists be required, procedures be established for using private and government tank trucks in disasters (for emergency water supply), fire departments adopt standard size hose couplings, measures be taken to integrate private communications networks into public emergency communications systems, support be given to establishing a federal requirement that local disaster plans be a condition of federal financial support, measures be taken to prevent the settlement of bridges and overpasses (due to liquefaction), anchoring and bracing be required of heavy equipment mounted on utility poles, regular independent review and testing be done of emergency response plans, an evaluation be done of the effectiveness of the state's Master Mutual Aid Agreement, local governments be required to address the seismic safety of parapet hazards, potential federal earthquake insurance be evaluated for its implications for California's market and risk, and that local governments be required to correct or eliminate potentially hazardous buildings.[145]

The JCSS' Executive Committee (chaired by Steinbrugge) and two of the advisory groups, Engineering Considerations and Earthquake Sciences (EC&ES) and Governmental Organization and Performance (GOP), dealt more than the other three groups with the successor organization issue. However, the Land Use Planning and the Post-Earthquake Recovery and Redevelopment advisory groups also contributed ideas from their perspectives while the Disaster Preparedness group (which I chaired) focused almost entirely on emergency preparedness and response issues.

There were strong differences of opinion between EC&ES and GOP, largely reflecting their membership's orientations: scientific and technical versus policy and governance. Several alternatives were hotly debated, but everyone understood that at some point the advisors would have to make one agreed-upon proposal to the Joint Committee before it expired in 1974. The pressure was on, especially after Governor Reagan appointed the GEC in 1972, because it might also consider some "successor body" before he left office on December 31, 1974.

Three models emerged from discussions within the advisory groups: A strong public commission, a strong mixed commission, and an advisory mixed commission.[146] The strong public commission "would have the power to set and enforce seismic safety standards. Commission members would primarily represent the 'general public' [and be advised by a technical committee—author]."

In addition, the debate included a "decentralized" model (regional bodies) or a "centralized" statewide organization with or without regional bodies. The headquarters location was debated: in a high-risk area such as Los Angeles or San Francisco, or in the capital city of Sacramento where legislation and budgets are decided. Each alternative and location had California precedents that could be drawn upon as the debate continued. In framing SB 1729, we made extensive use of the 1970 law (Chapter 891, Government Code) that established the Bay Area's Metropolitan Transportation Commission (for which I was serving as an Assistant Director with responsibility for legislative relations, among others).

During 1970–74 four other "relevant" legislative proposals were introduced, each of which could have precluded establishing an independent Seismic Safety Commission.

Senate Bill 530 was introduced on March 11, 1971, by Senator Tom Carrell of San Fernando. He was "a very powerful Southern California Democrat who owned a car dealership that sustained collapse of its building during the 1971 earthquake. That made a big difference, not just in his votes, but his influence, too. He wanted something done."[147]

"At the request of a seismologist of some repute," SB 530 called for creation of a "California Earthquake Office" with a director and staff appointed by the governor. It was referred on March 16, 1971, to the Senate's Governmental Organization Committee, where SB 530 was taken "under submission" (killed, for all practical purposes) nearly a year later on January 3, 1972. Senator Carrell's "Dear Al" letter said, "I'm sure it will never be revived this session." Senator Carrell expressed interest in the JCSS' work and a willingness to cooperate with Senator Alquist on seismic safety.[148]

The basis for future cooperation with the Reagan administration was being established when, speaking for Governor Reagan, the director of the Department of Water Resources, William R. Gianelli, wrote Senator Alquist, "inviting his attention to SB 530." Gianelli noted that "SB 530 is inappropriate at this time," and:

> We favor, instead, the plan of the Joint Committee on Seismic Safety which we understand is to take careful, detailed considerations of such an office or board along with their other considerations for recommended earthquake legislation. Services that the state requires, ongoing programs, capabilities of existing agencies to perform services, and needs for coordination would be considered.[149]

Another bill, Assembly Bill 1753, was introduced. It would have created an "Office of Earthquake Safety and Research." Speaking again for the Reagan administration, Gianelli repeated its concerns, noting that any new legislation should be consistent with the Joint Committee's findings.[150]

Both the JCSS and the GEC agreed that there would be no further efforts to legislatively create any organization until both groups had proceeded with their work and the JCSS had developed a conceptual proposal that could be a focus for negotiations. The final concept, an independent Seismic Safety Commission, was contained in Senate Bill 1729, which Senator Alquist introduced on February 14, 1974. Before I discuss the commission's charter in greater detail, however, we need to review the complex and lengthy internal JCSS and JCSS-Reagan administration negotiations that led to a mutually acceptable final version of SB 1729.

A "Public Safety Agency"

In California, multi-departmental "super agencies" have been created to "simplify" government operations. Each such agency is headed by a politically appointed Secretary, who has cabinet rank. One such proposal surfaced in 1973, which, had it passed, would have created a Public Safety Agency (PSA). On behalf of the Reagan administration, State Senator Robert Lagomarsino introduced Senate Bill 1152 on April 30, 1973. The June 22 version of the bill would have included the Department of the California Highway Patrol, Department of Corrections, Military Department, Department of Public Safety Services (combining the State Police and the Communications Division of the Department of General Services), Department of Youth Authority, Office of Criminal Justice Planning, Office of Emergency Services, Office of the State Fire Marshal, Peace Officer Standards and Training Commission, Adult Authority, Women's Board of Terms and Paroles, Narcotic Addict Evaluation Authority, and the California Crime Technological Research Foundation.

The proposed PSA would have been a natural home for a new Seismic Safety Commission. A GEC committee report would have had the PSA include a Seismic Safety Office (SSO) with an advisory Seismic Safety Commission. The office would have a small staff of about five people, and the Commission would have seven to nine members appointed by the governor. The proposal "would be used for discussion with the Joint Committee on Seismic Safety."[151] The SSO would be directed to:

102

1. Advise the governor through the secretary (of the Public Safety Agency) on earthquake-related matters.
2. Coordinate efforts at all levels of government and of the private sector in the preparation for and mitigation of earthquakes in California.
3. Serve as an information clearinghouse.
4. Channel ongoing and future earthquake loss reduction activities of all kinds through coordination, information sharing, and by providing seed money for programs and projects.
5. (Exercise) authority to review state agency budgets and grant proposals in earthquake-related activities and to advise the secretary thereon.
6. Review and advise the secretary on earthquake-related legislative proposals and to propose needed legislation to the secretary.[152]

A "Hazardous Buildings Commission"

Senator Alquist and his committee colleagues introduced twelve bills into the 1973–74 session, including SB 1729. Another of these was SB 2224, which was prepared by one of the JCSS' ad hoc committees because it observed that most local governments had not initiated hazardous buildings abatement programs in spite of Uniform Building Code sections that empowered them to order the abatement of such structures. Instead of a broad policy-focused advisory commission, SB 2224 called for a "State Hazardous Buildings Commission" having enforcement powers. Why?

There was clear agreement within the JCSS structure that existing earthquake-hazardous buildings were California's single biggest seismic safety issue. So, if SB 1729—the flag carrier of the JCSS' final legislative package—failed for whatever reasons, creation of an organization to at least address the hazardous buildings issue might be possible. If both passed, the technically oriented Hazardous Buildings

Commission would focus specifically on this single long-term issue. It remained possible to merge the two bills at almost any time during the legislative process to save one or the other.

Regional Planning and Governance: Organizational Models

The 1960s and 1970s were characterized nationwide by experiments in metropolitan/regional governance. Residents of the San Francisco Bay Area witnessed the creation of the Bay Conservation and Development Commission (BCDC), the Metropolitan Transportation Commission (MTC), the Association of Bay Area Governments (ABAG), the Bay Area Air Pollution Control District (BAAPCD), and others. A statewide commission established by initiative to regulate development of California's coastline and to ensure public access was not far behind.

Several legislative proposals also were introduced regularly to create a comprehensive multi-functional regional government agency. They always were controversial, and on one occasion "The powerful League of California Cities lambasted the regional government concept at their annual meeting in 1961, citing the 'home rule issue'."[153]

BCDC established its Engineering Criteria Review Board in 1968 to advise it on technical matters regarding construction along the bay's shorelines, including earthquake-related siting and design. This board also had members who served on the Resources Agency's two committees (see above), who later participated on the JCSS and the GEC, and who advised specialized state agencies on technical matters related to their particular missions: water resources facilities, dam safety, and freeway construction.

The San Fernando Earthquake Levels the Political Playing Field

Thanks to Senator Alquist's "adoption" of the seismic safety issue and an astute political strategy, the legislature's passage of SCR 128 in 1969 began prying open the proverbial window of opportunity for seismic safety policy advocates. However, the moderate 6.4 magnitude earthquake on February 9, 1971, immediately thrust open the window. For example, the day before (February 8) Senator Alquist was struggling to get some more money from the Joint Rules Committee to support the JCSS for 1971 via Senate Concurrent Resolution (SCR) 60. Joint Rules provided $10,000, but "the day of the earthquake, they were willing to give us any amount, half a million, a million, whatever" because, until the earthquake, legislative interest in seismic safety was low, and according to Steve Larson, "we on the Joint Committee were 'earthquake nuts'—just 'creepies'—those people who worry about things that only God can do anything about."[154]

Largely due to the Joint Committee's existence and abilities to move rapidly with the "advice" of its about 70 advisors, the cumulative policy impacts of this event are truly impressive, ultimately including the creation of the California Seismic Safety Commission—but we are getting ahead of the story.

Ronald Reagan Responds with the Governor's Earthquake Council

Melville Owen, a San Francisco attorney who served as a BCDC commissioner and who initially chaired the JCSS' Advisory Group on Governmental Organization and Performance (AG/GOP), also was a member of the Republican Central Committee. He tried to get Governor Reagan to intercede on behalf of SCR 60 but was unsuccessful, probably because this was an internal legislative matter.

In a 1973 speech before the California Engineering Foundation, Senator Alquist reflected on the pre-San Fernando earthquake lack of support, especially financial, for the JCSS' work. He recalled that following passage of SCR 128 "our problems had just begun [because] each year a funding bill must be passed." It (SCR 60) was approved unanimously in the Senate in 1970 only to be delayed in the Assembly "because of various reasons, none of which were related to the merits of the program." The session adjourned without action, leaving JCSS funding consideration for a hectic three-day veto session in late September.

Things sometimes change unexpectedly. Mel Owen mentioned "jokingly" at one AG/GOP meeting that the Republicans were meeting in San Diego the weekend before the September veto session, "and a small quake in the area on the weekend might do the trick." Alquist recounted that:

> Lo and behold, on the morning after the governor's arrival in San Diego a series of Richter scale 4 and 5 quakes hit the area. At breakfast that morning the governor was asked by Mr. Owen if he had felt the quake[s]. He responded, "Yes, indeed, it shook me out of bed." Three days later the funding bill passed the Assembly unanimously.[155]

Governor Reagan formed the Governor's Earthquake Council (GEC). The GEC existed from 1972 to 1974 and consisted of key appointed and career employees, a few of the Joint Committee's volunteer advisors, and others. Some state employees served on both groups, and this overlapping membership provided both a formal and informal communications channel that had enormous influence on "brokering" several pieces of seismic safety legislation sponsored by the JCSS and on budgets for several agencies' earthquake activities, such as adding $80,000 to the University of California's budget for the support of its seismographic stations.[156]

106

In appointing the 32-member Earthquake Council on February 2, 1971, Governor Reagan gathered "all state forces concerned with earthquake preparedness and research under a single banner…" and "he commended the legislature's Joint Committee on Seismic Safety for its 'excellent progress in the development of earthquake legislation'."[157]

Jean Laurin, formerly of the Seismic Safety Commission's staff, later observed that:

> If the creation of the Governor's Earthquake Council had any political implications, that fact is not evident from the historical files of either group and would be pure speculation. The ingredients are there, however. A powerful Democratic legislature had responded to the earthquake by creating a Joint Committee … and this body was already functioning when the San Fernando earthquake occurred…. Because of the high visibility of earthquake issues and the high level of public interest, it would make sense that a Republican Governor would want a similar body making recommendations to the executive branch.[158]

The political relationship between Republican Governor Reagan and Democratic Senator Alquist, who chaired the bipartisan JCSS, was not that cordial from the beginning. In addition to the "normal" executive-legislative tensions, there were partisan and philosophical differences, and moreover, Alquist had run against Reagan as the Democrats' candidate for Lieutenant Governor. This already tense relationship deteriorated further when Governor Reagan vetoed an Alquist-sponsored legislative amendment to the Office of Emergency Services' budget for funds to support the work of the JCSS' advisory groups (so the legislature would not have to use its money and to avoid a lack of internal legislative support by Senators Jack Schrade and Randolph Collier).

Commenting on the legislature's pre-San Fernando earthquake seismic safety policy leadership, Steve Larson reflected in 1989 that "We enjoyed that a lot because it proved that the legislature had beat[en] the governor to the punch."[159]

Eventually, this process allowed for negotiations that resulted in the passage and signing of SB 1729 in 1974: the Seismic Safety Commission Act. After the Joint Committee's four-year existence ended and as Governor Reagan finished his term, a "successor body" to both temporary organizations would be created.

The San Fernando earthquake forced both men to "rise above" these difficulties, at least as far as California's earthquake safety was concerned. The earthquake created a "political climate change" that led to the governor's willingness (with the help of insider supporters) to sign into law several Joint Committee bills, including the one establishing the Seismic Safety Commission.

Negotiating with the Reagan Administration Was Delicate

Karl Steinbrugge's dual membership provided a key link between the JCSS and the GEC so often informal negotiations about a successor body could proceed. As the process evolved, the JCSS' negotiating group was composed of Karl, me, and one or two others, and Governor Reagan's representatives James Stearns, Herbert Temple, and also one or two others. We met occasionally, usually for dinner, to prepare "mutually acceptable" language that was fed back to our respective constituencies, sometimes resulting in some "grumbling." On the JCSS' side, Steinbrugge took responsibility for the content of the legislation as it moved through and cleared the process, and Stearns did the same for Governor Reagan.

The Charter: Senate Bill 1729 of February 14, 1974

As might be expected, the Commission established by SB 1729 was a compromise that also accommodated GEC/state bureaucracy concerns so it would be acceptable to Governor Reagan. The Commission's initial members were to be drawn from both the GEC and the JCSS' advisors identified in lists supplied by each (and from which incoming Governor Brown reluctantly had to choose).

The Politics

The principle of legislatively directing a commission or some other organized unit of state government to attend to the state's earthquake threat had, as noted, been laid some years before SB 1729 was introduced. SB 1729's success in 1974 can be attributed to the concept's lengthy gestation period, during which significant negotiations about its final language could occur within the JCSS and between it and Governor Reagan's staff. This agreement in principle sent an always powerful bipartisan message to the legislature: Governor Reagan would sign the legislation, provided that the details could be worked out during SB 1729's deliberations. They were, and Governor Reagan signed SB 1279 on September 26, becoming Chapter 1413 of the Statutes of 1974. Overall, the legislation moved rather smoothly through the process with the real differences within the Joint Committee's advisors and the GEC's members already being negotiated between the two.

Getting Ready for the 1974 Session: Internal Drafting of the Commission Act

The overall strategy was to introduce a bill that was not unrealistic in its ambitions but that would contain several probably controversial points, which could provide room for compromises. The only principle that was not open to compromise was that the commission had to be independent (of all state government agencies). After several compromises were reached with the administration, it accepted the commission's independence. Had it not, Senator Alquist promised to drop the bill, and in the JCSS' and GEC's final year too many people had worked too long and too hard to see the effort die.

In December 1973, Karl Steinbrugge circulated a "concept draft" of what was to become SB 1729. Asking that it be kept "personal," he noted that the ideas came "largely from advisory group reports, supplemented by input from the Governor's Earthquake Council memos and information from other sources. It already contains compromises, and undoubtedly more will come."[160]

This concept draft proposed a nine-member commission, three each to be appointed from professional societies, universities, and the public, but attached to it to fuel the debate were five alternatives. The 13-member "Option A" provided for two public members and 11 others nominated by "established organizations": three from structural engineering, earthquake engineering, and soils engineering; three from a combined category of architecture and planning, fire protection, and local government; three from geology, fire protection, and public utilities; and two from insurance and social service.

"Option B" retained the two public members but broke down the remaining 11 members into individual categories: structural engineering, earthquake engineering, soils engineering, architecture, planning, geology, a full-time paid emergency services professional "employed by a city or county Office of Emergency Services"; a public utility senior manager nominated by the Public Utilities Commission;

110

and one each from insurance, banking, and manufacturing. "Option C" also provided for 13 commissioners, but six would be legislative appointees (not necessarily legislators themselves); "Option D" limited the Commission to seven members; and finally, "Option E" would have one-third of the commissioners represent the general public.

I assumed responsibility for preparing numerous versions of SB 1729, and many of the discussions about them occurred in the Metropolitan Transportation Commission's Berkeley offices. I also offered some thoughts on the Commission's composition:

> We are planning a public organization that will be responsible for public policy carrying forth well into the 21st century. The product of the Commission is *public policy*, which should be based on the best expert advice available. To my way of thinking, the Commission should be composed of public members [who are] experienced in earthquake problems, but by the very nature of their other experiences are aware of and sensitive to intergovernmental, public, and political externalities that might influence Commission policies.

> I like the mix of executive and legislative appointments and would like to see that a permanent part of the legislation. Such appointments do not preclude competent specialists and experts—selected for their own value as individuals, but it does separate the appointment of public officials from organizations which do not *have* to be responsive to the public interest. Such representatives might exercise an informal veto power over policies not acceptable to the profession but (would be) in the public welfare. Moreover, no appointments should be open to a specific agency ... since bureaucratic behavior being what it is, the agency probably will use its position to influence the regulator (commission) of some of its ... programs.[161]

Karl's revised version of January 4, 1974, (the Regular Session had just opened) was distributed on April 5, 1974, to an "ad hoc working group" of JCSS advisors. Among its most interesting provisions were that the proposed Seismic

Safety Commission might be included in another proposal to create a state Department of Building and Safety (an idea that surfaces periodically) or an interesting "last alternate": the Department of Finance itself.

What was the proposed Seismic Safety Commission going to do? Many of the proposed advisory responsibilities survived virtually intact to become law. Some of the proposed regulatory or organization-changing ideas never were included or were amended out as SB 1729 moved along. Several examples include:

"Reviewing and approving all state seismic safety standards and associated procedures prior to implementation by any state agency."

"Ordering compliance with state standards when, upon determination of the commission, non-compliance by any state, regional, or local public entity or district is evident."

"The commission, in conjunction with the Office of Intergovernmental Management, shall review all applications for state and federal funds to support earthquake-related projects and activities."

"In conjunction with the Department of Finance, review seismic safety programs and advise [on] state agency budgets."

[The commission] "Shall act as a review board after damaging earthquakes to assure that reconstruction measures reduce potential damage from future earthquakes."

"Monitoring periodic update of codes and insuring funding for future years."

"Selecting codes for lateral force which local [and] state jurisdictions must adopt (or just providing guidelines)."

The Processing (and the Versions)

JCSS Chairman Senator Alquist introduced SB 1729 on February 14, 1974. This started the formal process of creating an independent commission with limited regulatory powers, including transferring the responsibilities of the Strong Motion Instrumentation Board and the Building Safety Board to the Commission. This was done by repealing the applicable provisions of the laws establishing these programs. Both of these boards were established by early post-San Fernando legislation that was crafted largely by JCSS advisors (several of whom were to become initial commissioners).

Because of prescribed process deadlines, the strategy was to meet the Senate policy committee's first hearing deadline—which, if SB 1729 passed (and we were "guaranteed" that it would), would allow several weeks for intense negotiations before the bill's next hearing in Senate Finance. Thus, the most intense negotiations occurred during the spring of 1974.

The bill was amended several times during the next few months with Senator Alquist's and the Joint Committee's staff coordinating the process by seeking comments, drafting proposed amendments, reflecting concerns expressed by the Reagan administration and other interests, representing the ad hoc group's positions, and having numerous (sometimes several times a day) telephone conversations with those of us heavily involved in the process.

The bill's first stop was the Senate's Committee on Government Organization ("GO," not Governmental Efficiency), which was chaired by "friendly" Southern California Senator Ralph Dills. After author's amendments to the bill were accepted on April 15 (and it cleared its first policy committee), the bill sat for nearly 60 days (which allowed time for further discussions and negotiations. On June 13 Senate GO voted 7 to 1 to pass SB 1729 on to the Senate Committee on Finance. The amendments deleted the transfer of the Strong Motion Instrumentation Board and the Building Safety Board to the Commission but it did require each to "report annually" to it. These amendments also deleted the sections that would have allowed the Commission to be financed by the "unencumbered balances" of each program. These amendments eliminated two of the bureaucracy's major objections with the reporting requirement being an acceptable compromise.

One of the most important documents resulted from a June 11, 1974, meeting of the JCSS' ad hoc group handling SB 1729. It is a section-by-section marked-up copy of the April 15 version. Provided to Senator Alquist so it could serve as a basis for author's amendments at the June 13 hearing, it contains several important proposed amendments and marginal notes. The Department of General Services, Structural Engineers Association of Southern California (SEASC), and the California Legislative Council of Professional Engineers (CLCPE) proposed similar language to prevent the Commission from substituting itself for existing authorities and responsibilities (see below for wording).

Membership continued to be a major item for negotiation. In this version membership was increased from thirteen to fifteen based on proposals from the GEC and one of the JCSS' advisors. A proposal from the SEASC, the Structural Engineers Association of California (SEAOC), and the CLCPE would have specified establishment of a "technically oriented commission" by including 10 members from the architectural, engineering, and earth sciences professions (leaving room for only three to five others). This was not accepted by the ad hoc committee. Another proposed change, reflecting the tensions within the JCSS' advisors, would have commissioners be "principally representative of the public." One marginal note indicates that a "50/50" mix would be desirable.

The Portland Cement Company, which was represented on the GEC, proposed deleting the Commission's permissive ability to "accept grants, contributions, and appropriations from public agencies, private foundations, or individuals" because, according to the staff's notes, "it could create a contest among the various industries, product manufacturers and developers to donate funds in order to obtain favorable approval of systems, design criteria, and favoritism in disbursement of research funds." The company also objected to the prospective Commission being empowered to "do any and all other things necessary to carry out the provisions of this chapter" because as the marginal note indicates "it provides unlimited power to the Commission."[162]

Not unexpectedly, the Commission's proposed responsibilities also continued to be debated. Marginal notes indicate that comments were received on the April 15 version of SB 1729 from the ad hoc group itself, SEASC, SEAOC, CLCPE, Department of General Services, and the GEC. One staff note indicates concurrence with the advisory role and the Reagan administration's opposition to "bigger government": "This [version] will provide the Commission with wide-ranging authority to *implement by persuasion* new programs without the possibility of fiat. A super power semi-official agency is to be avoided *at all costs*."[163]

Many other relatively minor changes were noted, but perhaps politically the most important points were reserved for the staff's general comments at the end of the mark-up:

1. We have support statements from the Southern California Section, Association of Engineering Geologists, and East Bay Municipal Utility District.

2. The Department of Housing and Community Development might support the bill, but there is resistance to the creation of any new commissions.

3. The Department of General Services has informed us they will oppose the bill unless we include their amendments; SEASC will probably do the same."[164]

One (of my) proposed amendments, which was not accepted, would have increased the initial appropriations to $250,000 from $190,000. I had principal responsibility for estimating the costs for each version of SB 1729.

More author's amendments were accepted on June 20 in the Senate Committee on Finance. When bills reach the "money" committees (Finance in the Senate and Ways and Means in the Assembly) two new actors are heard from: the Department of Finance, representing the administration's position, and the Office of the Legislative Analyst. The bill remained with the committee until it narrowly passed 7 to 5 on August 6.

The close vote may be explained partially by the legislative analyst's concerns about "seemingly overlapping responsibilities" between the proposed Commission and the State Mining and Geology Board (and the Division of Mines and Geology) and that the cost "may be estimated at $40,000 per year, but may be much higher."[165]

So, what did the final June 20, 1974, version contain? Two of the legislative findings now read:

> Second, there is a pressing need to provide a consistent policy framework and a means for coordinating on a continuing basis the earthquake-related programs of agencies at all governmental levels and their relationships with elements of the private sector involved in practices important to seismic safety. This need is not being addressed by any continuing state government organization.

> Fourth, it is not the purpose of this chapter to transfer to the commission the authorities and responsibilities vested by law in state and local agencies.[166]

The first finding was our effort to distinguish the proposed Seismic Safety Commission from the Mining and Geology Board, and the second finding was added at the GEC's request so the bill would not attempt to "grab" any more programs.

The two provisions that would have empowered the Commission to ensure compliance with state standards were deleted, including an implementing provision that would have had the Commission "order compliance with state standards if, within a period of time determined by the Commission, noncompliance by any state agency or local government is shown."[167] However, new language was added directing the Commission to recommend changes to state agency standards "when new [scientific and technical] developments would promote seismic safety."

The amendments also increased the Commission's membership from 13 thirteen to 15 fifteen people by increasing the number of nominees made by the JCSS and the GEC from five to seven; deleted the provision that the chairman would be appointed by and serve at the pleasure of the governor, so that the Commission could select its own chairman; and shuffled the composition to include four local government representatives nominated by "established organizations," a seismologist, and someone with an electrical or mechanical engineering background. I recall that most of the membership changes were proposed by the GEC, especially the County Supervisors Association of California and the League of California Cities.

The amendments also granted the Commission authority to hold public hearings and to establish "working relationships" with any organizations "to further an effective seismic safety program for the state," and $190,000 was added to fund the Commission's first year (which amounted only to a recommendation to the fiscal committees).

En route to its second and third readings (on the Senate floor), SB 1729 was amended again before the Senate approved it 27 to 4 and sent it on to the Assembly on August 9. These amendments cut the funding to $85,000 because half of the fiscal year would be over on January 1, when the law took effect, and they specified the life of the Commission to end 61 days after the final adjournment of the 1977–78 legislative session.

Upon its arrival in the Assembly, SB 1729 was assigned to the Committee on Resources and Land Use. The bill was amended here, too, before the committee passed it on August 23 by a 5 to 0 vote to send it to the Committee on Ways and Means. The most important of these amendments was to add two members each nominated by the League of California Cities and the County Supervisors Association of California.

Ways and Means heard SB 1729 on August 28, and amended it again. The vote was 12 to 0. Most significantly, these final amendments shortened the Commission's life to 61 days after the final adjournment of the current 1975–76 session.

The Assembly passed the final version of SB 1729 by a vote of 68 to 0 on August 29, 1974. The Senate concurred with the Assembly's amendments, and SB 1729 was enrolled and was sent to Governor Reagan at 5:00 p.m. on September 5, 1974. Agreement having been reached, and supported by his key advisors, Governor Reagan signed the legislation on September 26, 1974. SB 1729 became law effective January 1, 1975, the day newly elected Governor Edmund G. (Jerry) Brown Jr. took office. Implementation started, therefore, with the arrival of a new and philosophically different Democratic governor.

The Result: Chapter 1413 of the Statutes of 1974

The best way to understand the final provisions of the Seismic Safety Commission Act of 1974 is to summarize the law. In addition to statements of findings about why the Commission was needed and how it emerged from the work of both the Governor's Earthquake Council and the legislature's Joint Committee on Seismic Safety, the act appropriated $85,000 to start the Commission; recognized that it was independent (a principle on which the JCSS would not compromise) but was advisory only; required it "to report annually to the governor and legislature on its findings, progress, and recommendations"; required the Strong Motion Instrumentation Board and the Building Safety Board to "report" annually to the Commission; and specified that unless "further legislative action" was taken, the Commission would last "only until the 61st day after the final adjournment of the 1975–76 legislative session" (about January 1, 1977).

The Commission's first seventeen members were to consist of fifteen appointed by the governor and confirmed by the Senate, and the speaker of the Assembly and the Senate Rules Committee each appointed a member. The new law prescribed staggered terms and rather tightly defined the initial and subsequent membership. Governor Jerry Brown had the "freedom" to appoint seven members from a list of sixteen nominees presented by the JCSS (from its group of volunteer advisors) and seven other members from a list of Reagan's GEC members.

Interestingly, the Commission's initial chairperson was to be appointed by the governor as long as the nominee had been "mutually agreed upon" by the JCSS' chairman, Senator Alquist, and the GEC's chairman, James G. Stearns, a Cabinet member and Governor Reagan's Secretary of Agriculture and Services. The nominee was Karl V. Steinbrugge. It is little wonder that Governor Brown, with virtually no control over the candidate pool, waited until mid–May 1975 to appoint the new commissioners.

The bipartisan nature of the JCSS and the cooperation between it and the GEC was reflected in the initial membership of the Seismic Safety Commission: eight Democrats, six Republicans, and one undeclared, plus the

WHEN GOOD SCIENCE WON

two legislative representatives. They were selected from a list of twenty nominees submitted by Senator Alquist on January 7, 1975, to incoming Governor Brown only days after his inauguration. Alquist noted that "all nominees are recommended by both groups as having superior qualifications to serve as commissioners."[168]

What about future members? We tried to constrain future appointments by specifying that members would be appointed from lists furnished by various established organizations: four from architecture and planning, fire protection, public utilities, and electrical and mechanical engineering; four from structural engineering, soils engineering, geology, and seismology; four more nominated by the League of California Cities and the County Supervisors Association of California; and the final three from insurance, social services, and emergency services. The legislature retained its two direct appointments.

What could this new and independent advisory Commission composed of volunteer members do to affect California's seismic safety? Not much from a regulatory perspective, but we depended on the expertise and experience of the Commission's members, our direct access to the legislature, and to cultivating cooperation with the agencies that were responsible for many seismic safety programs. In fact, a 1979 Commission study identified fifteen agencies that were giving attention to seismic safety within their program responsibilities.[169]

Briefly, Chapter 1413 authorized the Commission to set "goals and priorities"; request agencies to "devise criteria"; recommend "program changes"; review post-earthquake "reconstruction efforts"; gather, analyze, and disseminate information; encourage research; sponsor training; help "coordinate the seismic safety activities of government at all levels"; and establish and maintain "necessary working relationships ... to further an effective seismic safety program for the state."

The law gave the Seismic Safety Commission permission ("may" is the operative word in this context) to review earthquake-related state agency budgets and grant proposals and to advise the governor and legislature on them, but it excluded proposals flowing from California's universities to the federal government (e.g., National Science Foundation); permission to propose, review, and advise on legislation (one of the Commission's most influential activities); and permission to hold hearings on and recommend changes in state standards when "in the Commission's view, the existing situation creates an undue seismic hazard or when new developments [such as research] would promote seismic safety."

The two advisory boards (Strong Motion Instrumentation and Building Safety) that were required to report annually to the Commission were composed largely of former JCSS advisors who prepared the legislation establishing these programs soon after the 1971 San Fernando earthquake. One of the Commission's first activities was to establish committees of its own to examine the early implementation of other post-San Fernando JCSS legislation that many of its advisors helped prepare.

Immediate Challenges to the Commission's Existence

Implementation began with the swearing in of the first commissioners in May 1975. I resigned my appointment about 30 minutes later in the Governor's Office to take the position as the Seismic Safety Commission's first Executive Director. We opened the Commission's first offices across from the Capitol building on August 1, 1975. Sacramento was shaken by two earthquakes later that morning, and by about 3:00 p.m. several of us were in Oroville, about 70 miles north of Sacramento. This small rural community had suffered damage from two moderate events, which ultimately were to raise significant technical and policy issues about unreinforced masonry buildings, the design of the proposed federal Auburn Dam northeast of Sacramento, and the general seismicity of the Sierra Nevada foothills.

However, the euphoria of victory was short lived for the new Seismic Safety Commission: its survival quickly became **the** political challenge. First, the enabling legislation, Senate Bill 1729, gave the Commission a two-year life instead of four. This was a necessary deal made between Senator Alquist and Assemblyman Knox, both from the San Francisco Bay Area. While I never knew the details, Knox was prepared to hold SB 1729 in his committee (effectively killing the bill) unless Alquist agreed to support one of Knox's bills pending in the Senate. Alquist agreed but we had to settle for a two-year charter. Second, opening the office on October 1, 1975, (the day of the noted nearby Oroville earthquake) meant that I was well behind the normal state budget planning cycle for the next fiscal year (1976–77) that had begun in the spring.

While I had prepared budgets elsewhere, California's procedures were new to me, and I did not understand much of the Department of Finance's jargon. Assuming I would be contacted, I waited. Sure enough, I was contacted by the Department of Finance's assigned budget analyst, Calvin Smith, who needed a proposed budget "now." Calvin came to the Commission's offices and we prepared a budget in about an hour and a half. Problem solved. Third, and totally unexpectedly, recently elected Governor Jerry Brown created a "hit list" of boards and commissions he wanted to abolish in the interest of "streamlining" state government. The new Seismic Safety Commission was on the list.

Peter Drucker, the noted management scholar, said that management's principal responsibility is the organization's survival. What was the Commission's survival strategy? The following discussion of goals, tactics, and results are taken largely from my 102-page personal "survival journal" and supplemented by other supporting documents. Thus, the "politics of survival" became my principal focus for the first years of the Commission's existence.

These were intertwined foci encompassing legislative politics and tactics (again), including the Legislative Analyst's Office; bureaucratic politics, especially the Resources Agency's Department of Conservation and its Division of Mines and Geology (now the California Geological Survey) and its Mining and Geology Board; Governor Brown's office, its Office of Planning and Research, and his science advisor; the Department of Finance; and several other occasional key actors, including California's U.S. Senator Alan Cranston and Governor Roberto de la Madrid of the Mexican State of Baja California Norte.

The Commission's discussions of this overtly political strategy were the genesis of a longer running debate: to seek permanence or live with periodic "sunset" dates. Sunsetting programs was legislatively popular at the time. It meant that at specified intervals the legislature would review various programs to judge whether or not they should continue. I had mixed feelings about "to sunset or not to sunset," as did the Commission.

As time passed and my knowledge of legislative politics increased, I moved toward favoring (but never fully embracing) permanence because I feared that the Commission's future could be traded off for reasons that had nothing to do with earthquake safety. Last, Senator Alquist was there as our ardent and powerful defender, but it was not fair or politically smart for the Commission to become overly reliant on him because elections occur and life intervenes.

To address our survival, the Commission appointed a four-person "Strategy Committee" to work closely with me. Its members included Karl Steinbrugge, structural engineer and first chairman; Robert Rigney, current chairman and chief administrative officer for San Bernardino County; Louise Giersch, vice chairman and former mayor of Antioch and San Francisco Bay Area regional activist; and Professor Bruce Bolt, commissioner and noted UC Berkeley seismologist.

Our goals, simply stated, were to (1) legislatively extend the Commission's life to 1986 and to restore its budget for 1979–80; (2) defeat Governor Brown's proposal to eliminate the new Commission (along with several other boards and commissions) by December 31, 1979; and (3) to seize the initiative by seeking to expand the Commission's powers and responsibilities (i.e., "change the rules of the game"). Originally, we assumed that the Commission was going to have to address only goal number 1. We anticipated legislative success but feared a governor's veto. Now we were forced to confront the governor's formal effort to abolish the Commission. As the process unfolded, the actions we took sometimes addressed more than one goal. I was convinced that several key actors had a shared interest in seeing that the Commission be abolished for their own particular reasons.

We submitted the Commission's proposed budget for 1979–80. It showed a $90,000 reduction from the current one (1978–79) because we were going to complete a legislatively directed study at less cost than originally anticipated. Finance's assigned analyst called to say the department was very pleased and closed the item. Very soon thereafter, our analyst expressed anger when she learned that decisions to eliminate the Commission and cut its budget were made without any consultation with her, but she was the one who had to explain it to us, commenting that "I have heard that your budget has been reopened upstairs and I don't know what is being done with it." About then, her superior called me and asked if Senator Alquist was a member of the Commission. My "yes" ended the conversation. We saw the first official announcement in a *Sacramento Bee* article based on a reporter's comments after attending a legislative briefing on the governor's proposed budget. The bureaucracy has no secrets, but the victim often is the last to know!

One of my first tasks was to try to discover who was involved and what their motivations were. It appeared to me that the decision was made by small group of people including the Secretary of the Resources Agency, the Director of the Department of Finance, some assistants to Governor Brown, and the director and one or two staff members of the Governor's Office of Planning and Research (OPR). It also became clear that lower-echelon staff people, such as those in the Division of Mines and Geology and OPR, were not directly involved in this decision-making process. I believe that the secretary of Resources proposed abolition of the Seismic Safety Commission based on the contention that one of his departments, the California Division of Mines and Geology (CDMG), could perform the Commission's duties. This tactic would have been consistent with the attitude taken by a staff member of the Legislative Analyst's Office who was assigned to review the budgets of the Commission and the CDMG.

Legislative Tactics

The Seismic Safety Commission's normal "sunset" date was January 1, 1981. The Commissioners considered several alternatives: expire at the current sunset date, close at the governor's proposed date, extend itself for another five years, or seek permanent status by eliminating the sunset clause. The Commission decided to ask Senator Alquist and

Assemblyman James Keysor (from the San Fernando area) to introduce a bill to make the Commission permanent, with the idea that we would agree to a five-year extension, giving the Brown administration a fallback and face-saving position to extend the Commission's life until January 1, 1986.

Unexpectedly, the speaker of the Assembly introduced legislation to set sunset dates for the first time for a wide variety of other boards and commissions. This opened another possibility to extend the Commission's life. We monitored this bill's progress and were ready, if needed, to ask to be included in it, believing that the governor would not veto a bill that addressed far more than just the Seismic Safety Commission. The Strategy Committee thus tried to open up multiple avenues to optimize the Commission's chances for survival. This was intended to complicate the system.

The Commission also went on the offensive by raising the question of expanding both its regulatory and advisory powers. One was to legislatively seek the power of subpoena because the Commission had met some resistance from people who did not want to appear before it, and the Commission had heard of potentially helpful documents that it could not obtain voluntarily. The subpoena power was included in the survival legislation.

With some support from legislative staff members, the Commission's responsibilities might be expanded to include all natural hazards. This idea drew upon the earlier (1973) Division of Mines and Geology's *Urban Geology Master Plan*. It recommended that the duties of the group set up to oversee earthquake hazard reduction be expanded "to provide continuing cognizance over loss reduction programs for all geologic problems except the loss of mineral resources and flooding." This tactic was aimed directly at the Resources Agency by quoting one of its own reports.

126

Although Commission staff members drafted the language, and Assemblyman Terry Goggin from San Bernardino County offered to sponsor it, the Strategy Committee decided against pursuing it because some Commissioners, especially the "earthquake experts," saw this as a potential dilution of the Commission's focus and because the move might arouse the opposition of other state agencies that saw the Commission encroaching on their authorities.

Because several of the early Commissioners helped draft the Hospital Seismic Safety Act soon after the 1971 San Fernando earthquake, the Strategy Committee did approve recommending preparation of legislation to transfer the responsibilities of its advisory and appeals body (Building and Safety Board) to the Commission from the Department of Health (part of which later became the Office of Statewide Health Planning and Development [OSHPD]). The justification was that the advisory and appeals functions should be independent of the administering agency, and such functions would be consistent with the Commission's roles.

The Commission staff also was directed to draft a similar legislative proposal: to have it assigned the policy-guiding and appeals functions for the Special Studies Zones Act. This act, another early post-San Fernando measure written largely by advisors to the Joint Committee on Seismic Safety, is administered by the California Geological Survey (CGS, formerly the Division of Mines and Geology). Enacting this law would, similar to the above, provide independent appeals and advisory functions.

In the Commission's budget hearing in the Senate, Senator Alquist, a member of the Senate Finance Committee, said to the legislative analyst's representative (who remained concerned about the apparent duplication of functions), "well, that's fine; we are going to transfer those responsibilities to the Seismic Safety Commission, and I am introducing legislation to do it." After the hearing, during which our budget was approved, Senator Alquist turned to me and said "You'd better write those bills for me."

Independently evaluating predicted earthquakes also became a Commission survival topic. Earthquake prediction had become a "hot" scientific topic with strong and varied opinions about its feasibility being expressed by earth scientists, especially seismologists and geologists. The Director of the California Office of Emergency Services (OES) had administratively created the California Earthquake Prediction Evaluation Council (CEPEC), so the director had a screening mechanism to help determine the validity of predictions before asking the governor to issue declarations of emergencies. The Commission believed that the subject of earthquake predictions had become so publically visible that responsibility for evaluating their validity should be put into California law. However, while doing so might be good idea from a policy perspective, the Commission members feared that might bring OES into the Commission's survival battle, and it had not been a key actor up to now.

The long-standing Field Act that regulated public school construction, administered since 1933 by the state architect, led to the administrative creation of an appeals body. The appeals body used this committee to handle conflicts about public school construction regulations and administrative processes. The Commission's Strategy Committee considered drafting legislation to empower the Commission to provide these services to the Office of the State Architect so as to strengthen the principle of independent review and appeals (and to bring more responsibilities to the Commission). We did draft such legislation and we met with the state architect's staff to explain the possible benefits that the Commission saw in providing these services. The Commission's staff was instructed to refrain from pushing this legislation if it generated strong reactions because the Commission wanted to avoid mobilizing more opposition: the state architect.

Finally, the Commission's Strategy Committee considered two other survival possibilities. The first would have been to legally empower the Seismic Safety Commission to review and approve all seismic safety-related regulations developed by any other state agency. It would have meant that no state agency could issue any earthquake-related regulation before the Commission had reviewed and approved it. Such authority would have put the Commission directly into state regulatory processes. This idea was dropped because it almost certainly would arouse strong opposition, further complicate already complex administrative processes, and overwhelm the Commission with operational program responsibilities.

The second idea, also considered too complicated and potentially controversial to pursue, would have been to somehow redirect funds to the Commission through the budget process from the earthquake programs of other state agencies. Based on its priorities, the Commission would be empowered to allocate such funds among the programs and agencies so the state's overall approach to seismic safety would be more integrated.

Further Survival Considerations

The Commission's Strategy Committee briefly discussed the idea of the Commission assuming direct responsibility for managing programs. One possibility was moving the Strong Motion Instrumentation Program (SMIP) from the Division of Mines and Geology into the Seismic Safety Commission. Several of the Joint Committee's volunteers and new commissioners had written the original legislation and knew the program intimately. About fifteen to twenty staff positions and close to $1 million would come to the Commission. A second possibility was to move the

Structural Safety Section of the state architect's office to the Commission. About 100 staff positions and about $3 million would come to the Commission. Both ideas were scrapped because the Commission would become too operational and preoccupied with routine administrative, appeals, and regulatory operations.

The Strategy Committee also discussed funding for the Commission's operations. Our budget came from the highly competitive General Fund. One idea would have been to get legislative approval to allocate a percentage of the SMIP program's funds to support the Commission. This was a "special fund" program, the income for which comes from a small rate included in local building permit fees to cover the program's costs. The building industry was strongly opposed to this idea and labeled it as a "tax increase."

A second funding idea was to legislatively require the Commission to be allocated operating funds from the environmental license plate fund. This was appealing at the time because controversy had arisen about the administration of this several million dollar "slush fund" that was being used for "pet projects" of various kinds. Especially if the Commission's duties were expanded to include other natural hazards, it could argue that natural hazards risk reduction was an environmental protection measure.

The last idea was to seek funding for specific projects from the federal government, especially the new Federal Emergency Management Agency (FEMA), the U.S. Geological Survey (USGS), or the National Science Foundation (NSF). While the Commission did benefit later from occasional federal support, it feared creating a dependency and becoming preoccupied with federal priorities and grant and contract requirements.

Survival Plan Implementation

We employed three "in the trenches" tactics to implement the Commission's survival strategy: (1) influencing Governor Brown and other state agencies, (2) influencing the legislature, and (3) mobilizing outside support, including the news media. How the Commission and its staff implemented these interwoven tactics is summarized below.

Although Governor Jerry Brown and Senator Alfred Alquist were Democrats, this did not mean that they were friendly. How strongly this antagonism affected the Commission's survival is quite uncertain, but clearly the Commission was one of Alquist's "favorites." Regardless, the tactics we used to influence the governor and other state agencies included meetings with Governor Brown and key administration officials, asking our commissioners to make direct contact with officials and legislators they knew or felt comfortable contacting, and inviting Governor Brown's science advisor and former astronaut Russell Louis (Rusty) Schweickart to speak at the founding of the Western States Seismic Policy Council (WSSPC), which involved California's Commission, Utah's Seismic Safety Advisory Council, Nevada Governor O'Callaghan's Seismic Safety Advisory Committee, and invited representatives from other earthquake-prone western states.

We abandoned the idea of the Commission's chairman meeting personally with Governor Brown because the published budget reflected the governor's position. Quite surprisingly, Rusty Schweickart was interested in the Commission's work, and I believe he put in some "good words" for us during this process. Because the Commission was charged with advising the governor on seismic safety matters (and because we had no access of our own), I hoped to make the science advisor a bit dependent on us. His assistant began attending the Commission's meetings when the science advisor learned

131

we were taking the lead with the government of Mexico to examine earthquake risk associated with that part of the San Andreas Fault that is in Southern California and northern Mexico. We learned that this was a commitment made by the governor's office to Mexican officials "some time ago," but because of personnel changes the activity "fell through the cracks." In the latter 1970s, this led to the binational formation and funding of the U.S.–Mexico Earthquake Preparedness Project (USMEPP) modeled after the Southern California Earthquake Preparedness Project (SCEPP) and its San Francisco Bay Area counterpart, the Bay Area Regional Earthquake Preparedness Project (BAREPP).

The Director of OPR told me that the proposal to abolish the Commission "would not be a major campaign." Thus, our approach to the Director of OPR and the secretary of resources was softened. Our unachieved goal was to get the two leaders to request the Director of Finance to submit a "change letter" to the legislature indicating that their opposition was being withdrawn after "additional study of the situation."

I did meet with some Department of Finance staff and they were obviously uncomfortable. They talked in general terms about the need to reduce staffing, eliminate duplication of programs and services, and similar standard themes. Finance staff did go to great lengths to assure me that the reputation, competence, and expertise of the Commission and staff and the quality of our work was not an issue. By letting Finance's people talk, I did learn that the Seismic Safety Commission was "offered up" by the Director of the Department of Conservation and the Secretary of Resources in cahoots with a very small group of assistants to Governor Brown.

I also learned that within the Office of Planning and Research there was great conflict. One staff member with whom we worked regularly "went through the roof" when he saw the

Commission on the hit list. Our friend agreed to keep us informed of OPR's internal actions about the Commission's future. The OPR situation was complicated further by an earlier confrontation with an upper-level OPR staff member, who we practically threw out of the office because he accused us of taking state furniture out of the building. (We were taking it to the state's surplus property warehouse.)

This man was put in charge of preparing a confidential internal OPR paper about the Seismic Safety Commission. OPR requested some background information, which we assembled and delivered. Our hope was that the OPR staff would review the materials and become familiar with the Commission's work. Our OPR friend could not get a copy of this paper.

What tactics did we use to influence the legislature? Particularly because of Senator Alquist's continuing commitment to the Commission, and that the Assembly and Senate had seats on the Commission, and the fact that the law creating it charged the Commission with advising the legislature on seismic safety matters, we could operate rather freely and effectively even though the governor's office issued complex procedures and forms for dealing with legislators. Our direct access perhaps was our most valuable asset.

The Commission's first legislative goal was to delay our first hearing on the budget from early February to March so we had more time to build support. Rescheduling was easy to do, and in case problems arose, this tactic gave the Commission time to appeal or reopen the Commission's budget for further discussion in April, May, or June.

The second legislative goal was to introduce some Commission-sponsored legislation to give us visibility before various legislative committees, to demonstrate that we took our charge to advise the legislature very seriously, and to frighten (if not do permanent injury) to the Department of Conservation. We found authors in addition to Senator Alquist and the Commission's Assembly member to sponsor a bill extending the Commission's life; another to address special requirements governing the rehabilitation of earthquake hazardous buildings; a third to amend the Special Studies Zones Act; and a fourth to make minor changes to the Strong Motion Instrumentation Program. The last two are programs administered by the Division of Mines and Geology (now the California Geological Survey).

The Seismic Safety Commission members also became active in influencing the process. With example letters provided by staff members, many wrote their legislators. The Commission also appointed a small delegation to visit key legislators, but we discouraged personal visits because we learned from an earlier experience that some of the Commissioners were not effective in dealing with legislators. Thus, I concentrated on visiting Assembly and Senate members on the appropriations subcommittees that heard the Commission's proposed budget.

Assemblymember Terry Goggin from the San Bernardino County area became very helpful. Robert Rigney, then Commission chairman, was from San Bernardino County and they knew each other. Goggin helped us design a political strategy, expressed his anger at Governor Brown, and called the director of OPR to tell him that he would support continuation of the Commission in the Assembly. Terry Goggin also recommended that I act as a go-between for Senator Alquist and the governor because, as noted earlier, the two did not get along well.

Mr. Goggin also convened a subcommittee he chaired to review the Commission's work. These "oversight hearings" are commonly used to inform legislators about various issues. Unlike during budget hearings, personnel from the

Department of Finance and the Legislative Analyst's Office do not testify. Mr. Goggin said that since the proposal is to abolish the Seismic Safety Commission, "why don't you take whatever time you need to go through your written statement in detail." I read about 50 percent of it until the questions started. It turned out that this same subcommittee was assigned later to hear our budget. Mr. Goggin got the votes to restore the Commission's funds in the Assembly.

While seemingly unrelated, I learned that Mr. Goggin agreed to help Governor Brown locate authors for legislation the governor needed to have enacted to fulfill his budget wishes—like eliminating the Seismic Safety Commission. This gave Assemblymember Goggin great leverage on negotiating those bills that various legislators were willing to carry and those they were not. Not surprisingly, no legislator volunteered to sponsor the Commission's "death bill." Essentially, Governor Brown's legislative tactic was dead, but he could still veto the Seismic Safety Commission's budget.

Not surprisingly, the legislative analyst (again) recommended transferring the Seismic Safety Commission's responsibilities to the Department of Conservation and its Mining and Geology Board. The irony was that we had virtually no contact with the analyst's office during the year, but it issued a long analysis about duplication of functions. The analyst's report noted that the Department of Conservation could do all of the work by reassigning some of its staff. We took a wait-and-see attitude.

Surprise! The volunteer chairman of the Mining and Geology Board called the chairman of the Seismic Safety Commission (who in turn called me). The board chair expressed great concern that the two organizations were being "played one against the other" and that he did not trust the motives of the Department of Conservation's staff, which was working on a plan to absorb the Commission's responsibilities. The board chair took the position that the functions were complementary; the Commission's focus on earthquake risk was appropriate; the board was more technically oriented; that the board was not involved in "high-level" policy issues; and he saw the board's focus as helping the department administer its operating programs. Thus, the board chair proposed that a joint statement be prepared and signed to this effect and that the Commission staff prepare it. After minor tinkering, it was done and distributed to the "appropriate parties." I speculated that the chairman of the Mining and Geology Board might be one of the great unsung heroes of the Seismic Safety Commission's survival battle.

The third tactic was to mobilize external support for the Commission's renewal. It focused on three methods: (1) generating political support from outside of state government, (2) using personal contacts, and (3) contacting the news media.

Mobilizing political support from outside of state government was relatively easy. Dr. Frank Press, the President's Science Advisor at the time and well known to some commissioners, agreed to write to Governor Brown, as did U.S. Representative George Brown and California Senator Alan Cranston, both of whom sponsored the National Earthquake Hazards Reduction Act of 1977. The primary message was that abolishment of the Commission would seriously weaken the new federal program, cast doubt on the concept of shared responsibility between the states and the federal government, and undermine emerging seismic safety efforts in other states. In fact, Congressman Brown was said to have told Governor Brown that if the governor did not support the Commission, he (Representative Brown) would have a "very hard time" supporting Governor Brown's presidential campaign!

Further outside support was generated by our (Commission staff) contacting about 100 people who were on the Commission's committees and mailing list. We provided a letter from something akin to "Friends of the Commission," and asked each person to write to Governor Brown and the chairpersons of the Assembly Ways and Means and Senate Finance Committees urging that they support restoration of the Commission's budget. Commission staff did the work on our own time, and we paid for printing, postage, and other costs so we could not be accused of using state money for political purposes.

Additionally, the commissioners were asked to contact their peers and their technical and scientific organizations, such as the Earthquake Engineering Research Institute, to make contacts and send letters in support of the Commission's extension. The Commission included four local government members who asked their associations, such as the League of California Cities and the County Supervisors Association, to express their support because the Commission had streamlined some existing laws to the benefit of local government and were about to introduce legislation to allow local governments to adopt standards lower than those required for new buildings to facilitate the seismic rehabilitation of older structures. We also worked directly with the City of Los Angeles, League of Women Voters, and other organizations that had staff members or elected officials working with the Commission.

Commission staff members had deliberately cultivated relations with the print and broadcast media through providing monthly agendas, briefs on topics that might be of media interest, and occasional personal visits to the "Capitol News Bureau" offices near the Capitol Building. Building on these contacts, the Commission staff put together a press kit about the controversy over the Commission's survival and ensured commissioners that we were available for interviews. Several science writers were particularly supportive of retaining the Seismic Safety Commission.

Victory Again

Governor Brown signed SB 1729. The final version gave the Seismic Safety Commission an additional five-year existence. One of Governor Brown's staff members confidentially asked me to "call off the pressure (saying that) the Commission is off the hit list and not going to be eliminated." It would be my successors' responsibility to again address the question of permanence and the extended politics of organizational survival.

In Sum

Senator Alquist, who remained a politically savvy seismic safety champion throughout his legislative career, brought "outsider" seismic safety advocates into California's political system through the Joint Committee on Seismic Safety. Especially because of the 1971 San Fernando earthquake, the committee's influence increased dramatically, and the members' sponsorship of many hallmark laws was felt immediately. The senator's authorship of the legislation creating the temporary and later permanent Seismic Safety Commission "institutionalized" an advocacy coalition that continues to influence the state's political agenda.

Policy scholar Arnold Meltsner observed that even in California there was little widespread public support at the time for seismic safety. Specifically, regarding the JCSS' efforts he noted:

> Experts and their colleagues on the California legislature's Joint Committee on Seismic Safety, with skillful use of timing and, I would suggest, relying on an illusion of general public support, introduced and secured the passage of several important seismic safety bills.... In short, political support

for legislation came from small attentive publics of victims interested in *relief* and from a small attentive public of experts interested in *safety*.[170]

While the State Capitol building's complete restoration started with an almost secret letter about its earthquake vulnerability, it became one of the most expensive and challenging projects of its kind. The next chapter discusses only the Joint Committee's roles in helping launch the project.

ENDNOTES

110. Alfred E. Alquist, press release, Sacramento, October 23, 1969.

111. Notes and Comments, *Southwest Builder and Contractor,* June 10, 1933.

112. Advisory Group on Engineering Considerations and Earthquake Sciences (EC&ES), meeting agenda, Sacramento, February 11, 1971.

113. State of California Archives, "Records of the Joint Committee on Seismic Safety," Sacramento, n.d.

114. State of California Resources Agency, *Earthquake and Geologic Hazards in California: A Report to the Resources Agency*, Sacramento, 1967: p. I.

115. G. Jean Laurin, "SSC History and Commentary," unpublished personal communication, Sacramento, n.d.

116. Stanley Scott and Robert Olson, eds., *California's Earthquake Safety Policy: A Twentieth Anniversary Retrospective, 1969–1989* (Berkeley, CA: Earthquake Engineering Research Center, College of Engineering, University of California, Berkeley, December 1993): 13

117. State of California Resources Agency, *Earthquake and Geologic Hazards in California: A Report to the Resources Agency*, Sacramento, 1967: p. I.

118. Ibid., 14–15.

119. Ibid., 13.

120. Earthquake Engineering Research Institute, *George W. Housner, The EERI Oral History Series* (Oakland, CA: Earthquake Engineering Research Institute, 1997): 64.

121. Stanley Scott and Robert Olson, eds., *California's Earthquake Safety Policy: A Twentieth Anniversary Retrospective, 1969–1989* (Berkeley, CA: Earthquake Engineering Research Center, College of Engineering, University of California, Berkeley, December 1993): 9.

122. Ibid., 16.

123. Karl V. Steinbrugge, *Earthquake Hazard in the San Francisco Bay Area: A Continuing Problem in Public Policy* (Berkeley, CA: Institute of Governmental Studies, University of California, Berkeley, 1968): Foreword.

124. Stanley Scott and Robert Olson, eds., *California's Earthquake Safety Policy: A Twentieth Anniversary Retrospective, 1969–1989* (Berkeley, CA: Earthquake Engineering Research Center, College of Engineering, University of California, Berkeley, December 1993): xxiv.

125. Karl V. Steinbrugge, *Earthquake Hazard in the San Francisco Bay Area: A Continuing Problem in Public Policy* (Berkeley, CA: Institute of Governmental Studies, University of California, Berkeley, 1968): 63.

126. Ibid., 64.

127. Ibid.

128. University of California, Institute of Governmental Studies, *Public Affairs Report* 9, no. 6 (Berkeley, CA: University of California, Institute of Governmental Studies, December 1968).

129. Laurin, op. cit., 8.

130. Stanley Scott, in *California's Earthquake Safety Policy: A Twentieth Anniversary Retrospective, 1969–1989*, eds. Stanley Scott and Robert Olson (Berkeley, CA: Earthquake Engineering Research Center, College of Engineering, University of California, Berkeley, December 1993): 3.

131. Ibid., 8.

132. Ibid., 20.

133. L. Thomas Tobin, in *California's Earthquake Safety Policy: A Twentieth Anniversary Retrospective, 1969–1989*, eds. Stanley Scott and Robert Olson (Berkeley, CA: Earthquake Engineering Research Center, College of Engineering, University of California, Berkeley, December 1993): xi.

134. Karl V. Steinbrugge, "Earthquake Hazard Abatement and Land Use Planning: Directions Toward Solutions" (Geologic Hazards and Public Problems: Conference Proceedings, U.S. Office of Emergency Preparedness, Santa Rosa, CA, May 1969): 150.

135. Alfred E. Alquist, press release, Sacramento, October 23, 1969.

136. State of California Senate, *Daily Journal*, 5658, August 8, 1969.

137. Ibid.

138. Robert A. Olson, letter to L. Douglas DeNike, October 2, 1970.

139. Karl V. Steinbrugge, *Earthquake Hazard Abatement and Land Use Planning: Directions Toward Solutions*, op. cit.: 150–151.

140. Alfred E. Alquist, statement to the Advisory Group on Engineering Considerations and Earthquake Sciences (EC&ES), Sacramento, Februrary 11, 1971.

141. Advisory Group Coordinating Council, Sacramento, minutes of March 2, 1971.

142. EC&ES, agenda, Sacramento, February 11, 1971.

143. EC&ES, minutes, Sacramento, February 11, 1971.

144. Joint Committee on Seismic Safety, legislation proposals, Sacramento, March 11, 1971.

145. Ibid.

146. Anthony Jiga, Three Alternative Seismic Safety Commission Proposals, Joint Committee on Seismic Safety, Sacramento, May 24, 1974.

147. Louise Giersch, in *California's Earthquake Safety Policy: A Twentieth Anniversary Retrospective, 1969–1989*, eds. Stanley Scott and Robert Olson (Berkeley, CA: Earthquake Engineering Research Center, College of Engineering, University of California, Berkeley, December 1993): 7.

148. Tom Carrell, letter to Senator Alfred E. Alquist, May 24, 1971.

149. William R. Gianelli, letter to Senator Alfred E. Alquist, April 27, 1971.

150. Ibid. Letter to Senator Alfred E. Alquist, June 10, 1971.

151. James Stearns, Wesley Bruer, Herbert Temple, and Robert Jansen, memorandum: "Recommendation 26, GEC: Consideration of a Successor Body," Sacramento, July 3, 1973.

152. Ibid.

153. Laurin, op. cit, 3.

154. Steve Larson, in *California's Earthquake Safety Policy: A Twentieth Anniversary Retrospective, 1969–1989*, eds. Stanley Scott and Robert Olson (Berkeley, CA: Earthquake

Engineering Research Center, College of Engineering, University of California, Berkeley, December 1993): 6.

155. Alfred E. Alquist, "The Politics of Seismic Safety," speech to the California Engineering Foundation, Sacramento, December 9, 1973.

156. Bruce A. Bolt, in *California's Earthquake Safety Policy: A Twentieth Anniversary Retrospective, 1969–1989*, eds. Stanley Scott and Robert Olson (Berkeley, CA: Earthquake Engineering Research Center, College of Engineering, University of California, Berkeley, December 1993): 12.

157. Office of the Governor, press release no. 61, Sacramento, February 2, 1972.

158. Laurin, op. cit., 13.

159. Larson, op. cit., 7.

160. Karl V. Steinbrugge, cover memo, Sacramento, December 12, 1973.

161. Robert A. Olson, personal comments on Steinbrugge draft of January 4, 1974, Sacramento, n.d.

162. California Portland Cement Company, letter to Ken Ross, Los Angeles, April 19, 1974: 1

163. California Senate, Senate Bill 1729 (amended), Sacramento, April 15, 1974: 13.

164. Ibid., 17.

165. Legislative Analyst, "Analysis of Senate Bill 1729," June 11, 1974.

166. California Senate, Senate Bill 1729, Sacramento, June 20, 1974: 2.

167. Ibid., 6.

168. Alfred E. Alquist, letter to Governor Jerry Brown, Sacramento, January 7, 1975.

169. California Senate Office of Research, *Report on State Agency Programs for Seismic Safety,* Sacramento, June 4, 1979.

170. Arnold Meltsner, "The Communication of Scientific Information to the Wider Public: The Case of Seismology in California," *Minerva* xvii, no. 3 (London, UK, Autumn 1979).

WAVE TWO, CHAPTER 6:
The San Fernando Earthquake And Repercussions For The Capitol's Restoration

The West Wing of the State Capitol falls within a class of construction which suffers serious damage and is a collapse hazard in the event of a moderate local earthquake such as that which struck San Fernando on February 9, 1971.

Construction of California's Capitol started in 1860, but fires, floods, and the Civil War delayed its completion until 1874. In 1863, Governor Leland Stanford said "The State Capitol of California, that is to endure for generations, should be a structure that the future will be proud of, and surrounded with a beauty and luxuriousness that no other Capitol in the country could boast."[171] California's legislators moved into the new Capitol in 1869 even though work (primarily decorative) was not completed until 1874.

During subsequent decades the Capitol was remodeled and redecorated several times, leading one historian to make note of "well-meaning and misguided 'modernizations'." The major trigger was due directly to the change in the legislature from a part-time to a full-time body and the need for a permanent staff. For example, modifications to the historic building were made in 1955, 1957, 1960, 1962, 1964, and 1965. In 1964, the legislature began to address its longer-term space needs, which triggered a great debate about replacing the Capitol with a new twin-tower edifice or to restore and remodel the old Capitol building. Enter serious politics and the February 9, 1971, earthquake in San Fernando (365 miles south).

However, this was not the first time that questions were asked about the earthquake resistance of the State Capitol. It was well known that the April 21, 1892, earthquake in nearby Winters (31 miles west) damaged the Capitol. Although strongly felt in Sacramento, it is less clear that the 1906 San Francisco (88 miles southwest) or 1954 Dixie Valley, Nevada, (250 miles east) earthquakes added to the damage. The Dixie Valley event did, however, cause some damage to structures along the Sacramento River. The August 1, 1975, Oroville earthquake (70 miles north) also shook Sacramento, including the Capitol.[172]

It also is highly probable that San Francisco's 1906 earthquake (and fire) that nearly demolished the City Hall played a key role in influencing Speaker of the Assembly Leo McCarthy (from San Francisco) and his close colleague, neighboring San Mateo County Assemblyman Louis Papan (known as McCarthy's "enforcer"), to favor restoration of the Capitol (which became labeled as "Papan's Palace"). In fact, Speaker McCarthy "proclaimed that there would be no new Capitol. The original would be restored." Solidifying this statement, Assemblyman Leon Ralph and Senator Lou Cusanovich authored Assembly Bill 2071 and Senate Bill 1547 that called for the structural and aesthetic restoration of the Capitol. These bills were passed in 1975 and 1980 respectively."[173]

Researcher Stephen Tobriner chronicled the influential decision-making process that led to the restoration of San

146

Francisco's City Hall rather than its replacement. The old City Hall was severely damaged by an 1868 earthquake, and soon thereafter the legislature passed a bill calling for construction of a new City Hall on the site of a former cemetery on Market Street. The 1871 City Hall "was supposed to be earthquake-proof, and in the wake of the great Chicago fire of October 9, 1871, the (overseeing) commissioners insisted on making the building fireproof as well." For a variety of design and construction reasons, Tobriner notes that the new City Hall "was the proverbial disaster waiting to happen, and an exception to the real earthquake-resistant buildings of the time."[174]

The Joint Committee's Roles

The remainder of this chapter focuses on how the Joint Committee on Seismic Safety interacted with and influenced the policy decision to restore the Capitol, and how the new Seismic Safety Commission informally monitored the project as it progressed. Though very interesting, it is beyond the scope of this chapter to tell the fuller and very interesting story of the internal legislative politics of restoration (but it should be told). Suffice it to say that the legislature's Joint Rules Committee assumed jurisdiction over the restoration project and hired staff members, including John Worsley, the former state architect, to manage the project. It took about six years to complete at a cost greater than $68 million and was the biggest restoration project completed in North America at that time.[175]

Karl Steinbrugge reflected some concerns present in the earthquake engineering community about the Capitol's condition, and this concern resulted in a draft of the letter quoted above that he provided to Steve Larson, Senator Alquist's chief aide at the time. The draft became the senator's final request to the Joint Committee's advisors to conduct a very informal but candid evaluation of the Capitol's earthquake safety. Thus, three earthquake engineers (Karl Steinbrugge, Frank McClure, and Henry Degenkolb) and a preservation-oriented architect (George Simonds) examined the Capitol.

After the four circulated several drafts among themselves, they reported their evaluation in a letter sent *only* to Senator Alquist. In addition to the introductory note quoted above, the key portion of the letter said:

> As requested by you, the undersigned made a reconnaissance field inspection of the State Capitol on March 3, 1971.... The purpose of these efforts was to make a preliminary determination of the earthquake stability of the State Capitol in the event of a disastrous earthquake.... The earthquake resistance is provided by non-reinforced brick bearing walls.... The mortar is of poor quality by today's standards, and ... in one area, it had disintegrated to point where it could readily be removed by finger. [Thus,] the West Wing of the State Capitol is seriously deficient when compared to today's accepted minimum standards for earthquake safety ... and falls within a class of construction which suffers serious damage and is a collapse hazard in the event of a moderate local earthquake.... It is also our opinion that the upper dome and the rotunda probably represent the greatest hazard, and their failure— either vertically or laterally—could result in substantial number of injuries and life loss.[176]

In his cover letter to members of the legislature, Senator Alquist wrote that "any advice you might have about the steps that should be taken to ensure its safety would be appreciated."[177] Later, in a speech to the Western Conference on Disaster Preparedness, Senator Alquist injected a bit of humor noting that the team of experts "reported that the 102-year old structure was so weak, a moderate earthquake could bring the huge capitol dome crashing down on both houses of the legislature.... To my dismay, many people regarded this announcement as the best news out of Sacramento all year."[178]

Not long after this letter was filed, Emeritus Professor of Seismology Charles Richter (famous for the Richter scale), stated in a *Los Angeles Times* editorial on April 8, 1972, that: "There is quite a to-do in Sacramento now about the unsafe conditions of the State Capitol building. It is a serious matter, and of concern to all of us. I hope something can be done before there are serious consequences."[179]

Earlier Concerns

This was not the first time, however, that the earthquake resilience of the Capitol Building had been examined. John F. ("Jack") Meehan, a structural engineer with the State Architect's Office who specialized in the enforcement of the Field Act governing public school construction, provided some older file information to the EC&ES. These documents easily reinforced the recommendation of Karl, Henry, Frank and George to thoroughly investigate the earthquake resistance of the State Capitol. The documents included:

1. A May 25, 1928, *Memorandum Re(:) Alterations to Present Library Section of the State Capitol Building*. This inspection noted a number of structural deficiencies. The report states, "The structure as it stands would not stand an earthquake of any great intensity, but barring earthquakes will be perfectly safe and stable, providing no moving loads, such as assembly loads [i.e., meetings—author] or their equivalent are placed on the floors."

2. A Memorandum Reference to Room Used for Hearing by Senate Investigation Committee, Fourth Floor, State Capitol Building, Sacramento, California, dated January 17, 1933. This did not address earthquake forces, but it pointed out several deficiencies that led to making some repairs and ensuring the occupancy loads did not overload the floors' steel trusses. Of special interest was "structural design was not followed by the steel contractor in that not only were lighter sections used in numerous cases but cover plates and other

149

members called for were not furnished and the work was very indifferently done, with the result that this construction is not capable of carrying the load for which it was designed."

3. A *Building Vibration Report* done by the former U.S. Coast and Geodetic Survey in 1935 (two years after the Long Beach earthquake) and provided to Jack Meehan on January 29, 1971, just days before the February 9, 1971, San Fernando earthquake.

4. A *Notes of Trip* memorandum report dated April 22, 1949, by structural engineer Frank A. Johnson. This report addressed observations made following earthquake damage to the Washington State Capitol Building in Olympia. In comparing Washington's Capitol with California's, Johnson noted several deficiencies regarding the earthquake resistance of California's Capitol. They included undersized footings, poor quality mortar, and modifications that weakened the building. Johnson concluded "that the building as a whole as it now stands, has very doubtful resistance against earthquakes."

5. An October 19, 1949, Progress Report by Senior Structural Engineer Bruce M. Dack, titled *Preliminary (Seismic) Analysis of State Capitol at Sacramento in Connection with Proposed Rehabilitation,* pointing out several earthquake-related problems. They included weak mortar, questionable floor and roof anchorages, overstresses in the roof framing, "extremely poor" concrete quality, and non-conformance with approved alteration drawings.

6. An interdepartmental communication dated December 9, 1949, by Bruce M. Dack to D. C. Willett (also see Chapter 2), chief construction engineer, about the results of the Capitol's vibration tests. The memo said, "if the tests showed resonance between a portion of the structure and its supporting elements ... it would indicate a dangerous condition."

7. A memorandum report dated February 6, 1953, titled *State Capitol Seismic Analysis,* done by the staff of the then Division of Architecture, at the time a part

of the Department of Public Works. In the report, the staff assumed that the Capitol "as it now exists probably conforms essentially to the Riley Act ... which calls for a 2% seismic factor."[180]

8. The Department of Parks and Recreation's pictorial history of the restoration, noting that that "statuary decorated the Capitol's balustrade until 1906." The report observed, "At that time, most of the statues were removed in response to San Francisco's earthquake. Because of deterioration, the remaining statuary was taken down in 1948."[181]

Senate Concurrent Resolution 84 (SCR 84) was introduced by four senators, including Alquist, on May 24, 1971, and became the "trigger" for action.[182] The advisors to the Joint Committee on Seismic Safety, particularly the Advisory Group on Engineering Considerations and Earthquake Sciences (EC&ES), now became intimately involved in the legislature's decision-making about what to do about the West Wing of the State Capitol building.[183] At its meeting on November 11, 1971, the EC&ES voted unanimously to recommend that the state architect be chosen to do the evaluation and report with the assistance of an advisory board. This recommendation was given to Senator Alquist so he could feed it into the Joint Rules Committee's decision-making process.

After internal political processing, the *final* key section of SCR 84 reads: "That the Joint Rules Committee is hereby directed to determine whether the state architect or a private firm is best suited to evaluate and issue a report on the safety of the West Wing of the Capitol Building and determine the means and the probable cost of rehabilitation and/or reconstruction of the West Wing, including the alternatives of cost and method of eliminating the hazards, from earthquake standpoints, and shall contract with the party providing the highest quality of service at the lowest cost to the state for such evaluation, determination, and report, to be submitted to the legislature not later than March 1, 1972." The sum of $100,000 was allocated from the legislature's contingency funds to perform this work. Because of SCR 84, the legislature's Joint Rules Committee now became intimately and permanently in charge of the reconstruction of the Capitol Building, California's Bicentennial Project.

The EC&ES compiled a list of private firms deemed capable of conducting the study if the Joint Rules Committee chose to do that. However, on November 24, 1971, the Rules Committee decided to contract with the state architect. The JCSS also created from its members a four-person Study Committee on the Structural Status of the State Capitol (again Steinbrugge, Degenkolb, McClure, and Simonds). In a letter to Senator Alquist on June 14, 1971, the four committee members stated, "We believe that every effort should be made to reduce the use and occupancy of the West Wing."[184]

The state architect's report, *Seismic Study, West Wing, California State Capitol*, was submitted in June 1972. It offered six differing proposals with their estimated costs in 1972 values: (1) Continue Use (without correction, "very low"), (2) Rehabilitate (for present occupancy, $40,850,000), (3) Reconstruct (entire West Wing, $41,018,000), (4) Partially Strengthen (for full use, $15,003,000), (5) Vacate (and limit use, $2,074,000), and (6) Strengthen Portions (for museum use, $18,783,000). Where needed, these alternatives included costs to lease space in downtown locations for temporary legislative office space. These alternatives focused the internal legislative debates about what might be done.[185] The state architect also noted that if it (West Wing) is to be used for anything—monument or legislature—it has to be a building that meets all building codes and is a legal building in every sense of the word."[186]

John A. Blume and Associates, a prominent earthquake engineering firm, was hired to do a detailed evaluation and conceptual design for the Capitol's rehabilitation. In his oral history, John stated that "We found that even though the earthquake stability was not good, there were other conditions that were even worse. For example, the steel trusses over the Assembly rooms had been altered over the years and decades by tradesmen.... There were literally booby traps in that building that might have sprung, even without an earthquake.... The team of designers and builders saved all the artifacts and the exterior of structure, and essentially rebuilt the rest of the building inside-out."[187]

Follow-ups and Related Matters: Senate Concurrent Resolution 66 and Senate Constitutional Amendment 76

Included in these debates were questions about what, from legislative, public, and employee safety viewpoints, should be done *now*. In 1972 Senator Alquist introduced two legislative measures. The first was a Senate Concurrent Resolution (SCR 66) that would restrict the use of the West Wing, and the second was a Senate Constitutional Amendment (SCA 76) that would permit the legislature to exercise the powers granted to it in the time of an enemy-caused disaster to also be used in the time of a natural disaster.[188]

SCR 66 was held in the Senate Rules Committee ("killed") without further action in December 1972. However, the Joint Rules Committee did, in November 1972, prohibit tours of school children and posted signs that read "Visitors are advised that this building may not be a safe structure during earthquakes." The committee also ordered the evacuation of the Capitol building by June 1973, and Governor Ronald Reagan followed suit by directing administration officials to work with the legislature on the relocation.

To further complicate matters, powerful Senator Randolph Collier (not a friend of Senator Alquist's) was a champion of replacing the Capitol building with a new one ("Collier Towers"). He introduced Senate Concurrent Resolution 133 (SCR 133) in October of 1972. It established a Joint Committee on Legislative Building Space Needs. Collier asserted that the new Capitol could be done by 1977, and it would be in a new location, have had two 17-story towers over a 3-story base, and would cost about $70 million–$100 million. The fate of the original Capitol building would be left to the state government to decide. Senator Alquist, playing smart politics, supported Collier's call for a new Capitol **IF** Collier would support pursuing the state architect's proposal "F," calling for strengthening portions of old Capitol for use as a state museum.

Once the Seismic Safety Commission was operating (in the fall of 1975), it heard testimony from John Worsley on the restoration's progress, and it visited the Capitol several times during its restoration.

California's restored Capitol Building was reopened in January 1982.

Legacies of the Capitol's Restoration?

Just as San Francisco's restored City Hall helped trigger the restoration of the State Capitol building, is it possible that the restoration of California's Capitol triggered or contributed to similar other projects elsewhere? I was close enough to some of these projects and key actors to believe the following:

The Utah State Capitol: A seismic strengthening project was completed in 2008. It was retrofitted with a base isolation foundation system that would allow the building to survive motions from a 7.3 magnitude earthquake. The final cost for all work was $260,000,000.

154

Salt Lake City and County Administration Building: A major seismic rehabilitation was completed for this 1890s building, including using a base isolation retrofit method for its foundation.

The Nevada State Capitol: Its $6 million seismic strengthening and fireproofing was initiated by Governor Donal N. (Mike) O'Callaghan, my former federal boss who strongly supported my involvement in seismic safety, and who created a Governor's Advisory Committee on Seismic Safety that was supported by his science advisor from the Desert Research Institute.

Los Angeles City Hall: Councilmember Hal Bernson, who, as a young boy, experienced the 1952 Arvin-Tehachapi earthquakes in Kern County, led the effort to strengthen the City Hall. He was later appointed to the California Seismic Safety Commission. Also influential was Ezunial Burts, an assistant to Mayor Tom Bradley, who was on the Board of Directors for the Southern California Earthquake Preparedness Project (SCEPP). He later served on the Seismic Safety Commission.

ENDNOTES

171. Governor Leland Stanford, Annual Message, 1863, Sacramento, as quoted in *California State Capitol Restoration: A Pictorial History,* 1983.

172. On this day, the new Seismic Safety Commission opened its doors at 8:30 a.m. The building shook at about 10:00 a.m., and I was in Oroville at 3:00 p.m. to survey the impacts of the locally damaging earthquake. Other tenants in the office building expressed some lighthearted concern about having this new earthquake safety agency located there.

173. State of California, Department of Parks and Recreation, *Growth: Rebuild or Restore?* (Sacramento, CA: State of California, Department of Parks and Recreation, n.d.).

174. Stephen Tobriner, *Bracing for Disaster: Earthquake-Resistant Architecture and Engineering in San Francisco, 1838–1933* (Berkeley, CA: Heyday Books, 2006): 4:74–4:76.

175. Robert M. Wood, *Past Into Present: The Documentation of the California State Capitol House Museum* (Sacramento, CA: California Department of Parks and Recreation, July 1983): 1. The $68 million is considered by some to be a "soft" number because of the budgetary politics involved at the time.

176. Karl Steinbrugge, letter of May 3, 1971.

177. Alfred E. Alquist, letter of May 27, 1971.

178. Alfred E. Alquist, remarks, January 31, 1972: 18–19.

179. "The Quake Threat in Sacramento" (part 2 of 4-part series), *Los Angeles Times,* April 8, 1972.

180. California Division of Architecture, "State Capitol Seismic Analysis," interdepartmental communication, February 6, 1953: 1.

181. California Department of Parks and Recreation, *California State Capitol Restoration* (Sacramento, CA: California Department of Parks and Recreation, 1988): 22.

182. Resolutions express the will of the legislature. Concurrent resolutions must be agreed to by both the Assembly and the Senate before they become effective. They are not laws, and the governor's signature is not required.

183. The "West Wing" refers to the original Capitol, and the "East Wing" is an attached newer office building completed in the 1950s.

184. Steinbrugge, et al., letter of June 14, 1971.

185. Office of Architecture and Construction, "Seismic Study, West Wing, California State Capitol, Sacramento," June 1972: 51–100.

186. John A. Worsley, letter to Miss Brynn Kernaghan, Joint Committee on Seismic Safety, October 11, 1973.

187. "Connections: John A. Blume," *The EERI Oral History Series* 2, 1994: 84–85.

188. Constitutional amendments formally amend the state's Constitution and, therefore, become law.

WAVE TWO, CHAPTER 7: Protecting New Hospitals—A Tale Of Two Bills

We are struck that the big loss of life came in the collapse of two hospitals—one old and one new. By contrast, high-rise office buildings escaped with little damage. We cannot help but wonder if we are building offices better than we are building hospitals.[189]

Within two days after the February 9, 1971, San Fernando earthquake, and standing on the grounds of the newly dedicated but (fortunately) still empty and now failed Olive View Hospital buildings, I asked earthquake engineers Karl Steinbrugge and Henry Degenkolb, "Can't we do better in designing and constructing hospitals?" Unlike the nearby San Fernando Veterans Administration Hospital, where nearly all of the event's casualties occurred (49 of 58–65), the Olive View facility was new; the older unreinforced masonry VA hospital already was (and suspected by the VA to be) a disaster waiting to happen.[190]

Before Henry arrived and at Karl's direction, I remained outside as a lookout in case of aftershocks while Karl ventured in to study the severely damaged new main Olive View Hospital building. After his (thankfully safe) return and some quiet moments and discussion, the three of us agreed that something should be done to seismically protect

158

hospitals. The idea of the Hospital Seismic Safety Act was born, and we agreed to take the idea to our respective JCSS committees: Disaster Preparedness (DP) and Engineering Considerations and Earthquake Sciences (EC&ES), for further work on the concept. In retrospect, I consider the ultimate Hospital Seismic Safety Act to be second only to the 1933 Field Act that regulates public school construction.

I believed that the DP's members would be particularly interested in the losses of the emergency room and the entire ambulance fleet plus the inability of Olive View to accept incoming casualties. Understandably, Karl and Henry focused on the issues related to siting, design, standards, plan review, construction inspection, and related matters.

Preamble

On December 10, 1970, (immediately after the organization of the Joint Committee on Seismic Safety and about *two months before* the San Fernando earthquake), EC&ES met in San Francisco, where the members discussed seismic safety issues that the committee might address as it began its work. Hospital seismic safety was one subject, the minutes about which note that:

> It has been suggested that legislation be prepared requiring that hospitals (and perhaps all public buildings) be required to meet the Field Act (or some parallel standard). The problems concerning the building of hospitals in active fault zones should also be considered. How much hospital damage are we willing to risk?[191]

On January 14, 1971, (now about three weeks *before* the earthquake) EC&ES met again in San Francisco. One topic was hospital safety. Some the items noted for future examination were that hospitals should be made safe not only for those in the hospital already but also for those who will use the facilities after a disaster; hospitals must be capable of functioning immediately (after a disaster); these considerations should carry over to fire equipment facilities; and that the Department of Public Health reviews hospitals in all areas except construction, design review, and siting.[192]

The remainder of this chapter discusses the intent of the legislation; the principal policy and administrative issues we had to address in the legislation; the varied special interests that mobilized to influence the legislation; the legislative politics involved in Senate Bills 352 and 519; and key early implementation challenges.

Legislative Intent

I was responsible for drafting the intent of the proposed Hospital Seismic Safety Act. In addition to providing the guiding philosophy, the intent served to build consensus and to guide those involved with the complexities of the act's implementation. Little changed from the original; the final version of the intent from Senate Bill 519 reads:

> 15001. It is the intent of the legislature that hospitals, which house patients having less than the capacity of normally healthy persons to protect themselves, and which must be completely functional to perform all necessary services to the public after a disaster, shall be designed and constructed to resist, insofar as practicable, the forces generated by earthquakes. In order to accomplish this purpose the legislature intends to establish proper building standards for earthquake resistance based on current knowledge, and intends that procedures for the design and construction of hospitals be subjected to independent review.

Note that the intent was silent about what to require of existing hospitals, and note that it did not state—because we assumed it was clear—that the new law would preempt local code enforcement authority (and fees) governing the design and construction of new hospitals. In the first instance, we all recognized the ferocious battle that would come with requiring existing hospitals to comply with the new standards (but we hoped it might accelerate their replacement). In the second instance, it took a very early (i.e., "clean up") amendment to say that state preemption was one purpose of the new Hospital Seismic Safety Act.

Major Policy and Administrative Issues

Hospitals are different—very different in configuration, complexity, and occupancy—from public schools. Several major policy and administrative issues had to be addressed over two legislative sessions before the Hospital Seismic Safety Act became law.

In addition to adapting the Field Act's principles of construction standards, independent plan review and approval, and regular construction inspection, other policy issues included how the law would be administered (e.g., State Department of Health, Office of the State Architect, new agency); what the relationship (if any) would be with local building safety agencies; what the fee structure and mechanism would be for collection; what the distribution of responsibilities would be among the involved professions (e.g., architects, engineers, contractors, inspectors); whether an expert advisory and appeals body would be established; how buildings would be judged "completely functional" after earthquakes; whether penalties for nonconformance would be misdemeanors or felonies; and perhaps most conflictual, whether and by when the law would require existing hospitals to meet the requirements for new hospitals.

Senate Bill 352: Round One

On behalf of the Joint Committee on Seismic Safety, Senator Alquist introduced Senate Bill 352 on February 23, 1971. He noted that "in recognition of the vital need for the services of hospitals after an earthquake disaster, the intent of the (legislative) bill was to improve the capability of hospital structures to withstand the effects of strong earthquakes to the degree that they *would remain useable after earthquakes.*"[193]

This was only 14 days after the earthquake when the proverbial "window of opportunity" was wide open. We considered the hastily written SB 352 to be a "trial balloon," as it was designed primarily to find out what interests would become involved and what issues the legislation would raise. If it happened to pass and be signed into law, we felt that would be a "miracle." The second bill, SB 519, considered the points raised during the debates on SB 352, and we included acceptable changes to reduce potential conflicts during the following legislative session.[194]

Amazingly, internal work on the proposed draft of the legislation started on February 11 (two days after the earthquake) by an informal subcommittee of the EC&ES (chaired by Charles De Maria, a San Francisco structural engineer). Through the senator, and using EC&ES' proposal, the Office of the Legislative Counsel drafted Senate Bill 352 by February 23, and it was formally introduced ("read for the first time") that day, just two weeks after the earthquake. This subcommittee remained the focus of the hospital legislation, and its members spent enormous amounts of time handling comments, drafting amendments, attending meetings, providing briefings, and engaging in other activities involved in shepherding this legislation through the process.

SB 352's major initial provisions required plans for hospital construction and alteration to be made by structural engineers and licensed architects; established siting and construction standards (likely including for the first time potential fault rupture, estimated ground shaking severity, ground failure potential, and tsunami inundation risk); specified approval of plans and the conduct of building inspections to be performed by the Department of General Services

(Office of the State Architect—the enforcing authority for the Field Act); authorized the setting of fees to support the program, credited to a special revolving account; prescribed felony penalties for violations; and permitted the office to conduct periodic reviews of hospital operations to "assure [that] structural safety, elevators, standby equipment and emergency procedures, and procedures and facilities for storage of dangerous gases, liquids, and solids are adequate to resist earthquake tremor."[195]

Responses to SB 352

The bill did indeed flush out interested parties. Some were in favor, others were opposed unless it were amended to address their concerns, and some were totally opposed. Some of the interested parties included state agencies, such as the Department of Health and the Office of Architecture and Construction; the California Council of the American Institute of Architects; the California Trial Lawyers Association; the County of San Bernardino; the California Legislative Council of Professional Engineers; the City and County of San Francisco; the City of Los Angeles; the California Hospital Association; the League of California Cities; the Structural Engineers Association of California; the Construction Inspectors Association; the California Association of the Physically Handicapped; and other mobilized interests. EC&ES' subcommittee also dealt with comments submitted by other advisory groups, other EC&ES members, and professional colleagues.

A sample of the positions on SB 352 is reflected in the following quotes:

> Chairman Steinbrugge points out to those in attendance that he is opposed to the watering down of the bill. He would not be in a position to support an ineffective bill. Mr. Coogan of [the Department of] Public Health states that the bill is excellent

in concept. The [Division of the] State Architect agrees that DPH (Department of Public Health) should receive the plans and that the structural inspection should go to OAC (Office of Architecture and Construction). Mr. Steinbrugge adds that we must give the people of California confidence.[196]

The basic problems were administration of the bill, the financing of the structural safety review, the provision for felony as a consequence of the bill's violation, and the concept of the hospital building board.... Mr. Kimball ... wants the bill to encompass all health facilities (except state-administered hospitals).... The DPH would like the provision of Section 15021 to be amended to establish a structural safety committee under the present (Hospital Safety) Board.... Mr. Stolte, representing the CCAIA (California Council of the American Institute of Architects), states that their strongest objection has to be the liability present in Section 15016.... The possibility of getting insurance would be difficult.... Mr. Oppmann reports that the CHA (California Hospital Association) ... supports the concept of seismic safety in construction and operation as it relates to hospitals.[197]

Mr. Levy of The City and County of San Francisco's Department of Public Works opposed the draft of Senate Bill 352 because *"it preempts the field of plan checking and inspection of hospitals to the state.... Many jurisdictions have excellent building inspection departments that are completely capable of plan checking and of inspecting construction, and in some cases probably superior to state agency.... There is no reason for the state to preempt the authority of local jurisdictions."*[198]

The California State Employees Association, in a letter to Senator James R. Mills, president of the Senate, recalled that a June 30, 1970, progress report of the Joint Committee on Seismic Safety stated: *"Special consideration should be given to applying the successful aspects of Field Act construction regulation to those kinds of occupancies that will be critically needed after a strong or major earthquake (such as hospitals) or those containing large numbers of people."*[199]

Senator Alquist's amendments (only those recommended by the EC&ES after its review and made mostly in response to concerns expressed by interested parties) included requiring both public and private hospitals to comply, but not nursing and convalescent homes; designating the Department of General Services' Office of Architecture and Construction (Schoolhouse Section) as responsible for enforcing the structural safety of the hospital designs (via an interagency agreement with the controlling Department of Health); ensuring that only licensed architects, structural engineers, geotechnical engineers will be involved in plan reviews; adopting the same governing code as that adopted by the State Building Standards Commission; requiring continuous construction inspections; mandating certifications by design team members that the completed work was done according to the approved plans; and creating a Hospital Building Safety Board to "set policy, advise and act as a board of appeals." SB 352 was amended twice: on April 23 and May 25, 1971.

The Journey and the Death of SB 352

SB 352 began its legislative journey by first being heard by the Senate Committee on Health and Welfare on April 23, 1971. It was held for further consideration because of opposition and the promise by Senator Alquist to offer author's amendments. SB 352 was then heard again on May 25 but still held in committee because of more needed amendments. Finally, it passed the policy committee on June 3 and was referred to the Committee on Finance, where it was heard July 15, 1971. However, the bill was held in this committee, where it died without receiving a "do pass" vote that would have enabled it to move on to the Assembly.

SB 352 was killed because of Chairman Randolph Collier's enmity toward Senator Alquist. Senator Collier's power was so strong that no other committee members opposed

his motion to hold SB 352 in committee. This enmity had its roots in Senator Alquist's vocal opposition to Senator Collier's proposal to replace the Capitol building with a new "twin tower" edifice (which was never built) and other partisan and philosophical matters. This was a clear warning to us that we were likely to have the same problem when we tried again in the next legislative session, something Senator Alquist and I discussed as we left the hearing room.

Senate Bill 519: Round Two

On March 8, 1972, Senator Alquist introduced Senate Bill 519—essentially a reintroduction of SB 352—in the 1972–73 legislative session, where it was read for the first time and referred again to the Senate Committee on Health and Welfare (H&W). It was first heard there on April 13 and remained there because of the introduction of a set of substantive author's amendments. It was heard again on May 1 and then remained because of more substantive author's amendments. Finally, a week later, H&W passed the amended bill on to the infamous Committee on Finance.

SB 519 appeared before Senator Collier's Finance Committee on June 27, 1972. We feared the worst, but then a strange thing happened. After our testimony in support, Senator Collier recognized an unidentified man in the audience who said that the Hospital Seismic Safety Act might have become law earlier if Senator Collier had not held the bill in the Finance Committee where it died. Senator Collier replied by asking the man if he was criticizing how the chairman ran his committee, whereupon, almost without a breath and to demonstrate his power, Senator Collier moved that SB 519 be given a "do pass" vote. The members voted unanimously for the bill. A stunned Senator Alquist thanked Senator Collier, and we quickly left before any further comments could be made.

On July 3, SB 519 passed the Senate by a 28 to 2 vote, and was referred to the Assembly. It was heard on June 18 by the Assembly's Health Committee, where it passed unanimously with a minor (i.e., "nonsubstantive") amendment and with the recommendation to place the bill on the next committee's Consent Calendar (i.e., a list of "routine" measures where no opposition is expected, but where any committee member can request a bill's removal so it can be formally heard).

This next step was the Assembly's Ways and Means Committee ("Finance") where on August 3, 1972, SB 519 was passed onto the Assembly's Consent Calendar (because it was self-funding and did not have to compete with normal budget allocations). It was approved on August 4 by a vote of 71 to 0 and referred back to the Senate for concurrence on a minor Assembly amendment. The Senate concurred on November 9, and SB 519 was sent to Governor Reagan at 1:30 p.m. on November 14. With the internal support of the Governor's Earthquake Council[200] one week later, on November 21, 1972, Governor Reagan signed SB 519, upon which it became Chapter 1130 of California's Health and Safety Code and went into effect on January 1, 1973. News spread fast among the advisors about one of our greatest legislative achievements.

In the Hospital Act's case, such coordination proved useful. The GEC provided legislative language related to how the program could be implemented by the administrative agencies. For example, the Office of Statewide Health Planning and Development (OSHPD) would be the lead agency, and through interagency agreements it would have structural plan reviews done by the Office of Architecture and Construction and the State Fire Marshal. It also charged OSHPD with forming a Building Safety Board to help develop procedures, specifications, and standards as well as acting as an appeals body for applicants who had questions about or differences with OSHPD's findings.

In other cases, such as the legislation creating the Seismic Safety Commission, the GEC was less accommodating, which resulted in many amendments to weaken that legislation so that the Commission's powers would be limited to advisory functions and not intrude on agencies' authorities (i.e., "turf").

Based on many meetings, phone discussions, letters, and lots of volunteer time, SB 519 was similar to SB 352. The principal provisions included: directing the Department of Public Health to contract with the Department of General Services for plan review, construction inspection, and related services; requiring hospital owners to submit plans and pay fees into a revolving account; requiring the establishment of a Building Safety Board; prescribing penalties; and declaring "intent of the legislature to preempt from local jurisdictions the enforcement of building regulations adopted pursuant to this act, including plan checking."[201]

We all knew that the Hospital Seismic Safety Act's implementation would be very complicated[202] and might lead to "clarifying" amendments sometime later. This was to be very demanding work for the newly formed Building Safety Board, part of the Office of Statewide Health Planning and Development (OSHPD); the legislation to implement the law required the Building Safety Board to report annually to the Seismic Safety Commission.

A letter of December 17, 1973, to Senator Alquist from the Department of General Services demonstrated early actions being taken to implement the Hospital Act. It noted that DGS "is in full agreement with the intent expressed and will perform all duties imposed by both the Hospital Act (SB 519) and the Field Act.... We took preparatory steps to write regulations and propose budget revisions as soon as the bill was signed by the governor."

Recognizing hospitals owned by the federal government, such as the Department of Veterans Affairs, Senator Alquist wrote the administrator in the "hope that your office would consider requiring similar if not the same standards for veterans hospitals built in California."[203]

One major and unexpected issue rose quickly: double charging hospitals for plan reviews and building inspections. While the stated intent of the Hospital Seismic Safety Act was to preempt local authorities related to hospital design and construction (as the Field Act did for schools), clearly, some local governments continued to charge hospital owners for plan reviews and inspections. In addition to contending with two sets of fees, hospital owners would face sometimes conflicting requirements. With strong support from the California Hospital Association (CHA), an early notice to local governments and hospital owners helped bring an end to this practice.

Confronting Later Issues

Controversies over the new Hospital Seismic Safety Act began almost immediately, and I believe an entire chapter could be written just about them. Some had to be dealt with legislatively, and many more became agenda items for the newly created engineering-oriented Building Safety Board, the regulation-making and appeals body.

Among the key items were standards and methods for securing nonstructural items, such as piping, medical equipment, supplies, and furnishings; regulations applicable to other on-campus non-medical buildings, such as power and heating/air conditioning plants and storage buildings; multiple inspections and the credentials of the inspectors;

and conflicts between local versus state regulations and inspections.

Other matters also had to be addressed. Examples include the act's applicability to medical facilities other than hospitals (e.g., outpatient surgical centers, mental health facilities, psychiatric hospitals, skilled nursing facilities); excluding wood frame buildings, such as small rural hospitals; additions or extensive modifications to existing hospital buildings; and the requirements for new hospital construction in areas with lesser seismic activity, such as the Central Valley.

Drawing on experiences with the Field Act, we understood that time would be needed for the engineering community (including the state's plan review and inspection staffs) to develop the expertise and experience to effectively address the construction of new hospitals. We also acknowledged that many, many hours would be needed to negotiate particulars with interested stakeholders like hospital owners and their design teams, the California Hospital Association, and many others.

In a category by itself was the issue of what to do about non-complying existing hospitals, a subject we deliberately avoided in crafting the original legislation. Retroactively changing the rules "after the game has been played" always is a major legal, regulatory, administrative, financial, and political challenge. Again, another entire story could be written about this one, but we understood that this would have to be addressed by others later. During my tenure with the Commission, I did have occasional "chats" about this subject with key staff members in the Office of Statewide Health Planning and Development (OSHPD). We talked about a 50-year compliance period that would have to be enacted legislatively. The 50 years would allow, for example, time for detailed engineering evaluations, setting applicable standards, deciding about retrofitting or replacing hospitals, incorporating new medical technologies and their needs, financial planning, and deciding about differing requirements for large urban hospitals compared to small rural ones.

Following the 1994 Northridge earthquake, Senate Bill 1953 required evaluations to be done and then structural and nonstructural upgrades to be made to existing acute care hospital buildings over 30 years. The costs of compliance raised significant concerns.

ENDNOTES

189. KNX radio editorial, Los Angeles, February 10, 1971

190. Within months after the earthquake, a group of patients' relatives sued the VA in federal court for negligence. A VA engineer (James Lefter) produced a typewritten memo reporting that the hospital complex was vulnerable to earthquakes and that steps should be taken to replace or strengthen the buildings. The presiding judge ruled against the plaintiffs when he said that while the VA was aware of the risk and was taking positive action to reduce it, it was not the VA's fault that the earthquake occurred when it did.

191. Advisory Group on Engineering Considerations and Earthquake Sciences, Sacramento, December 10, 1970: 3.

192. Advisory Group on Engineering and Earthquake Sciences, Sacramento, minutes of January 14, 1971: 3.

193. Alfred E. Alquist, letter to Lawrence R. Robinson, Director, Department of General Services, Sacramento, January 10, 1972. Emphasis added.

194. I quickly learned earlier that there is a strong preference toward resolving conflicts before a bill is formally heard in committee. That way, open conflict (objection) is greatly reduced, making it easier for members to vote and to avoid tying up time with lengthy discussions. Thus, "no opposition" is a welcome announcement by all interested parties.

195. Senate Bill 352, Legislative Counsel, Sacramento, February 23, 1971.

196. Advisory Group on Engineering Considerations and Earthquake Sciences, Sacramento, minutes of May 13, 1971.

197. Ibid., Sacramento, minutes of June 10, 1971.

198. Robert C. Levy, letter to R. J. Diridon, consultant to the Joint Committee on Seismic Safety, San Jose, CA, March 5, 1971: 1.

199. Loren V. Smith, letter to James R. Mills, Sacramento, February 11, 1971: 1.

200. Soon after the San Fernando earthquake, Governor Reagan appointed a Governor's Earthquake Council (GEC). It was established partly in response to the Joint Committee's lead in the legislature. The GEC focused on what might done by the executive branch to improve seismic safety. In spite of the disdain between Reagan and Alquist, the GEC's chairman, James Stearns, reached out to the JCSS to establish coordination. The result was overlapping membership that included some state agency representatives (e.g., John Meehan, EC&ES and Office of Architecture and Construction, and Karl Steinbrugge, "CEO" of the JCSS' Advisory Groups).

201. Legislative Counsel, Senate Bill 519, Chapter 1130, Health and Safety Code, November 21, 1972.

202. Lawrence R. Robinson Jr., letter to Alfred E. Alquist, Sacramento, December 17, 1973.

203. Alfred E., Alquist, letter to Donald Johnson, Sacramento, February 22, 1973.

WAVE TWO, CHAPTER 8:
Other Laws Triggered By San Fernando

I am still astonished at the progress we have made in 20 years. I think that is not true of almost any other discipline.... On the other hand, you can't get too impatient. It takes time to change things.[204]

I wrote the previous chapters to "give life" to California's legislative process so readers can learn about what it takes to go from a public policy concept to its enactment. This chapter briefly highlights some other subjects to round out the story.

Other Joint Committee Legislative Achievements

While I chose only a few successful legislative examples of the Joint Committee's work to illustrate the process involved, there were others. They include:

- Enacting the Strong-Motion Instrumentation Program. This bill directed the then Division of Mines and Geology to organize, purchase, install, and monitor instruments in representative structures and geologic environments. An early amendment was passed to permit the division to purchase the requisite data processing equipment because the original legislation did not specifically authorize it.

173

- Passing the Alquist-Priolo Geologic Hazards Zones Act. This bill expanded the membership of the State Mining and Geology Board, and instructed that body to prepare policies and criteria for the development of designated special studies zones that include major active fault traces. Additional fees were authorized to charge applicants for building permits for sites within such zones, with the revenues to be split by the state and local jurisdictions. The state geologist was directed to prepare maps of the zones for use by local and state agencies.

- Enacting a requirement that local building departments maintain as public records plans of buildings for which the department has issued a building permit. This was especially important for post-earthquake studies so engineers could evaluate damage in light of the buildings' original design and specifications.

- Passing legislation requiring owners of dams designated by the California Office of Emergency Services to prepare dam failure flood inundation maps for use in planning evacuations. Local agencies affected by this law were required to prepare emergency evacuation plans and procedures. This law was based on a City of Los Angeles program that proved instrumental in identifying the area below the nearly failed Lower Van Norman Dam, collapse of which might have involved evacuating an 80,000-person area in the designated flood plain.

Not Always Successful: Some Legislative Failures

While the Joint Committee on Seismic Safety's policy efforts helped put California into a position of global prominence to reduce earthquake risk, the story would be incomplete if we omitted failed legislative attempts.

Many forces affect the legislative process, some of which are illustrated in the previous chapters. The negative ones

174

often simply are the flip sides of the positive forces. These include a legislator's personal relationships with his or her peers; informal and formal expressions of opposition by various interests (and campaign contributors); inability to meet internal legislative process deadlines; and decisions by the Joint Committee's volunteers to "pull" the legislation for further work (due usually to internal differing opinions about content).

Rarely did legislative committee actions formally "kill" bills as a result of our failure to secure enough "aye" votes. Sometimes we knew we were in trouble when other "coded" messages told Senator Alquist that a bill was floundering. Examples that almost always got unanimous committee "aye" votes included referring a bill to "interim study" (which meant nothing); holding the bill in committee pending the submission of "author's amendments" (commonly done if the author agreed with the proposed amendments, or if not, Senator Alquist just let the bill die); asking the author to withdraw the bill so further work could be done on it; and once in a while, a chair who was a friend of Senator Alquist would advise him against bringing a particular legislative bill to the committee in question ("Al, you don't have the votes today").

These contingencies often led Senator Alquist's staff to try persuading the clerks in both Rules Committees' offices to assign bills to friendlier committees and not to the "graveyard" committees or those chaired by one of Senator Alquist's "enemies."

Committee chairs also could direct their staff to leave a hearing's agenda open for the remainder of the day. This allowed us to "round up" friendly absentee legislators' votes by asking them to call committee staffs to record their "aye" votes. At the end of the day, we would call committee offices to learn of the final votes. One other tactic over which we had virtually no influence would be for a committee member to quietly change his or her vote after a bill was heard in a committee hearing. When such changes were done after the committee adjourned, its staff would quietly record the change. Only at day's end would we know the final results. We always hoped for a margin of two or three "aye" votes to reduce this uncertainty.

Another tactic, available only during Year 1 of a two-year session, would be to "convert" a bill to a two-year bill. This allowed holding the bill over for further work and negotiations until Year 2. Then in Year 2 it would be heard again before the committee that held it over.

Some examples of failed Joint Committee legislation included:

Senate Bill 897 (1972), Assessment of Geologically Hazardous Lands

This bill would have established procedures whereby property owners could obtain reduced property assessments based on geologic reports submitted to county assessors, providing the property owner agreed to refrain from developing such property for the areas determined to be unsafe.

Senate Constitutional Amendment 42 (1972), Assessment Valuation

This bill, held for interim study, was a companion to SB 897. By amending California's (lengthy) constitution, it would have allowed the legislature to provide for the reassessment of damaged property after the lien date for a given year.

Senate Constitutional Amendment 76 (1972), Legislature's Disaster Powers

SCA 76 was held in the Assembly Rules Committee. The legislation would have extended to natural disasters the situations in which the legislature could convene and fill the offices of deceased members.

Senate Concurrent Resolution 66 (1972), Capitol Visiting Restrictions

Held in the Senate Rules Committee, SCR 66 would have prohibited guided tours of the West Wing of the Capitol, prohibited entrance to the wing by persons under 18 without a waiver, and required posting of warning signs of structural danger at entrances.

Senate Bill 424 (1973), Assessment Valuation

This bill was held by Senator Alquist after its first hearing before the Senate Revenue and Taxation Committee. Like SB 897 of 1972, it would have provided for the immediate reassessment of damaged property provided its use is "appropriately revised."

Senate Bill 1372 (1973), Future Emergency Service Structures

SB 1372 was scheduled to be heard in January 1974 before the Senate Committee on Governmental Organization. The bill provided for the development of construction regulations for future appeals procedures and would have been enforced by local agencies.

Senate Bill 1374 (1973), Existing Emergency Service Structures

This bill was held by Senator Alquist in the Senate Committee on Governmental Organization. It would have required bringing existing emergency service structures up to code when funds became available from state and local bonds (see SB 1372).

Senate Bill 1375 (1973), New Equipment in Emergency Service Structures

Held by the author in the Senate Committee on Governmental Organization, and the final bill in the "Emergency Service Structures" series, the legislation would have forbidden the installation of new federal or state-funded communications and/or disaster equipment in structures not meeting the standards that would have been established by SB 1372.

Senate Constitutional Amendment 13 (1973), Assessment Valuation

Held in the Senate Revenue and Taxation Committee, this bill, a companion to SB 424, would have permitted the legislature to establish regulations regarding the assessment of damaged property.

The New Seismic Safety Commission Takes the Baton

The new Seismic Safety Commission opened its doors in Sacramento on October 1, 1975. Among its early agenda items were continuing to work on some final Joint Committee legislation, discussing some minor "cleanup" legislation to clarify language in earlier bills, and evaluating the implementation status of selected earthquake safety laws. The Commission members believed strongly in their responsibility to "advise the legislature and governor" on earthquake safety matters.

ENDNOTE

204. Stanley Scott and Robert A. Olson, eds., *California's Earthquake Safety Policy: A Twentieth Anniversary Retrospective, 1969–1989* (Berkeley, CA: College of Engineering, University of California, Berkeley, December 1993):90.

Part III. Waves Three and Four—Loma Prieta and Northridge

WAVES THREE AND FOUR, CHAPTER 9: Waiting For The Next Wave— Only An Intermission

Earthquakes remain a threat to California as the Pacific and North American plates continue to move and we all wait for the "big one," not to mention the far more frequent locally damaging events. This chapter summarizes key legislation proposed and enacted after the 1989 Loma Prieta ("Wave Three") and 1994 Northridge ("Wave Four") earthquakes. Data came from a December 2000 California Senate Office of Research report, *A History of the California Seismic Safety Commission.*[205]

I have not discussed these in detail because, while I stayed in the hazard mitigation/emergency management field, I was not directly involved with legislation following either of these events. However, I hope readers appreciate the full scope of how state earthquake safety policy has evolved since 1933 and the full array of the subjects that California's legislature has addressed over the decades.

Wave Three: The October 17, 1989, Loma Prieta ("World Series") Earthquake

This 6.9 magnitude event caused 67 deaths and about $5 billion in damages. In response, the legislature's 1989–1990

Regular Session (including a parallel Extraordinary Session following the earthquake) saw the enactment of bills:

- permitting the California Housing Insurance Fund to guarantee loans for specified earthquake safety improvements;

- specifying that income from a ¼ percent temporary increase in the sales and use tax be allocated for response and recovery costs;

- requiring the Seismic Safety Commission and the Office of the State Architect to develop a policy on acceptable levels of earthquake risk for new and existing state-owned buildings;

- to create a $300 million bond act to repair, reconstruct, replace, relocate, or retrofit state and local government buildings;

- mandating surgical clinics to meet specified earthquake standards for licensure;

- requiring local agencies to review design and construction of local bridges;

- directing the Seismic Safety Commission to hold an Earthquake Research Evaluation Conference; and

- transferring $80 million from a special fund to the San Francisco-Oakland Bay Bridge and I-880 Cypress Structure Disaster Fund.

In addition to these, legislation also was passed to:

- request the president to work with state and local agencies on a coordinated intergovernmental search-and-rescue response plan;

- extend the duration and funding for the Bay Area Regional Earthquake Preparedness Project and its Southern California counterpart;

- require the Department of Transportation to inventory all state-owned bridges that require strengthening or replacement;

- require the State Mining and Geology Board to develop guidelines for the preparation of maps of seismic hazard zones and develop priorities for such mapping; and

- authorize redevelopment agencies to determine when structural repairs to specified and historic buildings are necessary in order to meet various earthquake safety codes and standards.

Forty-five more laws were enacted during the same 1989–1990 session.

Wave Four: The January 17, 1994, Northridge Earthquake

This Southern California 6.7 magnitude earthquake killed 57 people and caused up to $44 billion in damages. California's legislature responded by calling another parallel Extraordinary Session. Some illustrative examples of passed legislation include:

- a $2 billion bond act to repair, replace, reconstruct, or retrofit transportation infrastructure, schools, hospitals, utilities, and other facilities and to finance housing repair loans;

- revising permit requirements related to repairing, replacing, or retrofitting various local facilities;

- making substantial changes to the Hospital Seismic Safety Act by requiring the enforcing agency (Office of Statewide Health Planning and Development) to review and approve schedules, compliance documents, and construction documents related to damaged facilities;

- creating the California Earthquake Authority to develop and implement a residential earthquake insurance program that includes hazard mitigation measures;

- requiring local enforcement agencies to inspect mobile home parks every eight years;

- requiring creation of a funding formula to retrofit the San Diego-Coronado Bridge;

- making changes in the Seismic Safety Commission's membership; and

- requiring application of provisions of the Field Act to relocatable buildings leased or purchased before or after specified dates.

Further measures clearly stemming from the Northridge experience were enacted through the 1999–2000 Regular Session.

Since then ...

Little has happened legislatively since the Northridge earthquake. Based on new technologies, in 2016 a measure was enacted (Senate Bill 438) that created the California Earthquake Early Warning System, and since the early 2000s lawmakers introduced several bills seeking to relax the Hospital Seismic Safety Act's standards. Moreover, the Seismic Safety Commission lost its cherished independence when it became part of the Business, Consumer Services, and Housing Agency; and in 2020 Governor Newsom's budget for 2020-21 included a proposal to relocate the Seismic Safety Commission to the Office of Emergency Services. The battle for independence never seems won.

However, I am reminded of geographer W. M. Davis' closing sentence in his 1934 article on the Long Beach earthquake: "Life is full of hazards, and we must take our chances among them. The chances of an enjoyable life in Southern California are, in spite of its occasional earthquakes, undeniably excellent."[206]

ENDNOTE Chapter 9

205. Seismic Safety Commission, *Living Where the Earth Shakes: A History of the California Seismic Safety Commission* (Sacramento, California Senate Office of Research, December 2000. The Commission later published this document as SSC 2000-04.

206. W. M. Davis, "The Long Beach Earthquake," *The Geographical Review* XXIV, no. 1 (January 1934): 11.

Part IV. References, the Author, Index

SELECTED REFERENCES

I list here publications that had a direct bearing on composing this book. It is not a complete list of the many more documents, such as meeting minutes, news articles, periodicals, and drafts of legislation that I examined. The listed documents include several that provided background information and influenced my thinking about scope, methods, style, and subjects that I believed were important to this history. These citations are in addition to many of those listed in the endnotes for each chapter. With a few exceptions, I chose not to repeat those here.

Bruce A. Bolt and Richard H. Jahns, *California's Earthquake Hazard: A Reassessment*, Public Affairs Report, 20, no. 4 (Berkeley, CA: Institute of Governmental Studies, University of California, Berkeley, August 1979).

California Seismic Safety Commission, *Living Where the Earth Shakes: A History of the California Seismic Safety Commission* (Sacramento, CA: California Seismic Safety Commission, December 2000).

W. M. Davis, "The Long Beach Earthquake," *The Geographical Review* XXIV, no. 1 (American Geographical Society of New York, January 1934).

Carl Henry Geschwind, *Earthquakes and Their Interpretation: The Campaign for Seismic Safety in California, 1906–1933* (Ann Arbor, MI: UMI Dissertation Services, 1996).

David B. Hattis, John F. Meehan, and Donald K. Jephcott, *Seismic Mitigation Strategies for Existing School Buildings Which Are Subject to Earthquakes Throughout the United States* (Silver Spring, MD: Building Technology Inc.,1993).

Joint Committee on Seismic Safety, *Meeting the Earthquake Challenge: Final Report to the Legislature, State of California, by the Joint Committee on Seismic Safety,* Sacramento, California. (Reprinted by the California Geological Survey as Special Publication 45, January 1974).

Joint Technical Committee on Earthquake Protection, *Earthquake Hazard and Earthquake Protection* (Los Angeles: Los Angeles Chamber of Commerce, June 1933).

Arthur E. Mann, *The Field Act and California Schools: A Report to the Seismic Safety Commission,* Sacramento, CA, March 1979).

Peter J. May, *Disaster Policy Implementation: Managing Programs under Shared Governance* (New York: Plenum Press, 1986).

Richard Stuart Olson, Robert A. Olson, and Vincent T. Gawronski, *Some Buildings Just Can't Dance: Politics, Life Safety, and Disaster* (Stamford, CT: JAI Press Inc, 1999).

Robert A. Olson, "Legislative Politics and Seismic Safety: California's Early Years and the 'Field Act,' 1925–1933," *Earthquake Spectra* 19, no. 1 (February 2003).

Robert A. Olson, "Technology and Public Policy for Earthquake Safety," in *Structural Engineering and Structural Mechanics,* ed. Karl S. Pister (Englewood Cliffs, NJ: Prentice-Hall Inc., 1980).

Robert Olson Associates Inc., *To Save Lives and Protect Property: A Policy Assessment of Federal Earthquake Activities, 1964–1987* (prepared for the Federal Emergency Management Agency, Sacramento, CA, 1988).

Stanley Scott and Robert A. Olson, eds., *California's Earthquake Safety Policy: A Twentieth Anniversary Retrospective, 1969–1989* (Berkeley, CA: College of Engineering, University of California, Berkeley, December 1993).

Stanley Scott, *Policies for Seismic Safety: Elements of a State Governmental Program* (Berkeley, CA: Institute of Governmental Studies, University of California, Berkeley, 1979).

Karl V. Steinbrugge, *Earthquakes, Volcanoes, and Tsunamis: An Anatomy of Hazards* (New York: Skandia America Group, 1982).

Karl V. Steinbrugge and Carl B. Johnson, "Earthquake Hazard and Public Policy in California," *Engineering Issues: Journal of Professional Activities* (New York: American Society of Civil Engineers, October 1973).

Syracuse Research Corporation, *The Role of States in Earthquake and Natural Hazard Innovation at the Local Level: A Decision-Making Study* (Syracuse, NY: Science and Technology Center, December 1984).

U.S. Office of Emergency Preparedness Region 7, *Geologic Hazards and Public Problems: Conference Proceedings*, eds. Robert A. Olson and Mildred M. Wallace (Santa Rosa, CA: U.S. Office of Emergency Preparedness Region 7, May 1969).

THE AUTHOR

Robert A. Olson is the retired president of Robert Olson Associates Inc., a California-based consulting firm that specialized in hazard mitigation, disaster prevention, emergency management planning and training, contingency planning, policy research and advocacy, and community post-disaster recovery planning. Mr. Olson earned a BA in political science from the University of California, Berkeley, in 1960 and his MA in political science from the University of Oregon in 1964.

He served as (1) Planning Officer and Regional Representative for the U.S. Office of Emergency Preparedness (now the Federal Emergency Management Agency) from 1964 to 1971; (2) Assistant Director, San Francisco Bay Area Metropolitan Transportation Commission (MTC) from 1971 to 1975; and (3) the first Executive Eirector of the California Seismic Safety Commission from 1975 to 1981. His consultancy remained open from 1981 until his retirement, with occasional project work continuing until 2020.

During his career Mr. Olson has been affiliated with UC Berkeley's Institute of Governmental Studies; Earthquake Engineering Research Center; and the Center for Environmental Design Research. While with the MTC, he led the organizing of the BART Impact Studies program between MTC and UC Berkeley's Center for Urban and Regional Development.

Mr. Olson also has held research positions at Stanford University, University of Illinois, California Institute of Technology, Florida International University, and the University of Southern California. In addition, he has independently reviewed hazards research proposals and projects for the Oak Ridge National Laboratory's Associated Universities.

He was the first member with a social science education admitted to membership in 1973 in the Earthquake Engineering Research Institute. He served as its Vice President, a member of the Board of Directors, as Acting Executive Director, a member of several post-earthquake investigation teams, and a member of the Editorial Board, and served on various committees, including Public Policy, Urban Earthquake Hazards Reduction, and the Seismic Performance of Buildings. Honorary membership was bestowed on him in 2004.

Mr. Olson helped design and oversee the initial development and testing of FEMA's "HAZUS" multi-hazard loss estimation software. He was a contributing author to the National Research Council's book titled *Facing Hazards and Disasters,* and wrote an article on a federal mitigation law's implementation for the *Journal of Hazard Mitigation and Risk Assessment.* He co-authored *Some Buildings Just Can't Dance: Politics, Life Safety, and Disaster* that examined the problems the City of Oakland had after the 1989 Loma Prieta earthquake. He and others completed a research study on the building safety problems that the City of Oroville in Butte County faced following the 1975 earthquake. Mr. Olson was a member of California's Tsunami Policy Working Group and worked with a consultancy to prepare a guidebook for local planners on minimizing exposure to potential tsunami damage.